Vernon Morwood

Wonderful animals

Working, domestic and wild

Vernon Morwood

Wonderful animals
Working, domestic and wild

ISBN/EAN: 9783337228590

Printed in Europe, USA, Canada, Australia, Japan

Cover: Foto ©berggeist007 / pixelio.de

More available books at **www.hansebooks.com**

WONDERFUL ANIMALS:

WORKING, DOMESTIC, AND WILD.

Their Structure, Habits, Homes, and Uses.

DESCRIPTIVE, ANECDOTICAL, AND AMUSING.

BY

VERNON S. MORWOOD,

AUTHOR OF 'FACTS AND PHASES OF ANIMAL LIFE,'
AND LECTURER TO THE ROYAL SOCIETY FOR THE PREVENTION OF CRUELTY TO ANIMALS.

'So from the first eternal order ran,
And creature linked to creature, man to man.'
POPE.

'To me be Nature's volume broad displayed.'
THOMSON.

WITH EIGHTY-ONE WOOD ENGRAVINGS.

LONDON:
JOHN HOGG, PATERNOSTER ROW.
1883.

PREFACE.

—∞°∞°∞—

S this book is intended by the author to be a
companion volume to 'Facts and Phases of
Animal Life,' he has referred particularly, and
at some length, to the Working Domestic Animals, and
others not treated upon in his former volume.

In the present work the author has introduced very many
Amusing and Interesting Anecdotes never before printed,
which refer to the Instinct, Structure, Habits, Uses, Intelligence, and other things characteristic of Our Animals,
both Wild and Tame.

An earnest endeavour has been made to guard both old
and. young against the attendant evils and demoralizing

influences arising from an indulgence in Acts of Cruelty to Animals, and to induce them to cultivate a Kindly Feeling towards Our Dumb Companions. To treat animals humanely will help to make them even more willingly useful as servants of man, and be as the bright sunshine of their existence.

V. S. M. ·

CONTENTS.

—◆—

CHAPTER I.

Curious Odds and Ends about Animals.

CHAPTER II.

Peeps down a Microscope.

CHAPTER III.

Lilliputian Subjects of the Animal Kingdom.

Contents.

CHAPTER XXII.

Friends of Animals.

GLOSSARY

EXPLANATORY OF WORDS USED IN THIS VOLUME.

———•———

Accipitres—Birds of prey.
Amphibious—That which partakes of two natures, so as to live in air or water.
Animalculæ—Animals of the very smallest kind.
Anseres—Web-footed or aquatic birds, as the goose.
Antennæ—The feelers of insects.
Apterous—Insects without wings.
Aquatic—Animals and plants that live and grow in water.
Articulated—Having joints, but no internal skeleton, as lobsters, crabs, etc.
Biped—Having two feet only, as men and birds.
Bivalve—Animals having two shells to open and shut.
Bombyces—Moths.
Bovidæ—Hollow-horned ruminating animals, as cows, oxen, etc.
Byssus—A tuft of silky filaments which issue from the shell of some kinds of mollusca
Carnivorous—Feeding on flesh.
Cartilage—A gristly substance, found in the noses of burrowing animals.
Caudal—Belonging to the tail, especially in fish.
Cheiroptera—Having wings like hands.
Chrysalids—Pupæ of butterflies.
Cocoon—A soft covering spun by caterpillars.
Coleoptera—Shield-winged insects.
Conirostres—Birds having beaks of conical shape.
Cranium—The skull.
Crustacea—Some kinds of shell-fish, as shrimps, prawns, crabs, etc.
Crust—The hoof, or outside covering of the horse's foot.
Cygnet—A young swan.
Diptera—Insects with two wings, and whose mouths are formed for suction only.
Dorsal—As the muscles on the back of the horse, and the fins on that of a fish.
Edentata—Being destitute of teeth.
Elytra—The wing-sheaths of beetles.
Embryo—A bud, or first rudiments of an animal.
Ephemera—A short-lived insect ; an evening fly.

Equidæ—The horse family.
Felt, and Felting—Cloth made of wool without weaving ; a compact sub-stance of fur and wool rolled and pressed together in lees or size.
Fluviatile—Living in rivers, streams, or ponds.
Gallinaceæ—Birds resembling farmyard fowls.
Gallinæ—An order of birds including peacocks, pheasants, turkeys, etc.
Genus—A class of animals including many species.
Glandular—Having reference to the glands of animal bodies, which are smooth fleshy substances.
Glires—An order of animals comprising mice, field-mice, beavers, hares, rats, jerboas, etc.
Gormand—An animal that eats ravenously, as a pig.
Grallæ—Wading birds, as the avocet, etc.
Gyrations—A circular motion, or whirling round.
Habitat—The home, or place of resort, of an animal.
Halcyon—Serene, quiet, peaceful, etc.
Haw—A gristly substance growing under the nether eyelid and eye of a horse.
Hemiptera—Insects with wings partly leathery and membranous.
Herbivorous—Living upon herbs and vegetables.
Homoptera—Insects with wings all alike.
Hybernate--To pass the winter in some secluded place, without leaving it to seek either food or water ; as bats, hedgehogs, etc.
Hybrid—Mongrel, produced by the mixture of two species.
Hymenoptera—Insects with membranous wings.
Imago—The last and perfect state of an insect.
Incubation—Sitting upon eggs to hatch the young.
Infusoria—Microscopic animalculæ found in liquids of various kinds, and even in rain, snow, and fog.
Insectivora—Birds and quadrupeds that live on insects.
Instinct—A psychical property with which all animals are endowed, prompt-ing them to defend themselves, to procure food, and to perpetuate their own kind.
Invertebrata—Animals that have no backbone.
Lachrymal—Generating tears.
Lamina--A sensitive and soft substance lying between the coffin-bone and the inside of a horse's hoof.
Larynx—The upper part of the windpipe.
Larvæ—Insects in the caterpillar, or grub state.
Lateral—Growing out on the side.
Lepidoptera—Insects with scaly wings.
Lilliputian—A person of very small size.
Ligament—A substance which binds one bone to another.
Mallard—The drake of the wild duck.
Mammalia—Animals that suckle their young.
Mandible—The upper or lower jaw. The instrument of manducation.
Membrane—A thin, white, flexible skin which serves to cover some part of the body.
Metamorphosis—Change of shape, as of a chrysalis into a winged animal.
Migration—The act of removing from one country to another. Many kinds of birds are migratory.
Milt—The soft roe of fishes.
Muzzle—The mouth of anything, as of a dog or gun.
Natatory—Enabling to swim. Having special reference to birds with webbed feet.
Nervures—The corneous divisions in the wings of insects.
Neuroptera—Insects with four transparent wings containing many nerves.

Nocturnal—Animals that are concealed by day, but which come out by night in search of food, etc.

Nymph—Another name of chrysalis. The second stage of an insect's life.

Olfactory—Pertaining to the sense of smelling.

Omnivorous—All-devouring, or eating all kinds of food.

Organism—Having reference to organs or natural instruments, as the tongue, the organ of speech, etc.

Ornithology—A minute description of birds.

Orthoptera—Comprising straight-winged insects.

Oviparous—Laying eggs.

Ovipositor—The organ of insects by which eggs are deposited.

Ovis—The genus to which sheep belong, etc.

Pachydermatous—Animals having very thick skin, as the elephant, the rhinoceros, and the pig.

Parasites—Insects that live upon other animals.

Pectoral—Pertaining to the breast, or to the fore-fins of a fish.

Phlebotomy—To let blood from a vein.

Phosphorescent—Shining with a faint light which emanates from phosphorus, a substance of a yellowish colour.

Physiology—The science which treats of the structure and constitution of animal bodies.

Pupa—The chrysalis, or quiescent state of an insect.

Raptorial—Birds that are carnivorous and live upon other birds and small quadrupeds.

Rodents—Animals whose teeth are formed for gnawing.

Rotifera—Wheel-bearers.

Ruminating—Chewing the cud, as cows and sheep, etc.

Saliva—An excretion from certain glands of the mouth which moistens the food and assists digestion.

Sallender—A rough horny patch situated in front of the hock of a horse.

Scansores—Birds that are organized for climbing.

Scintillation—The act of sparkling.

Serrated—Notched on the edge like a saw.

Solidungulate—A mammal with a solid hoof on each foot, as the horse, donkey, etc.

Spawn—The eggs of fish and frogs.

Succedaneum—A substitute, or that which is used as something else.

Tentacles—Feelers, as in the sea anemone.

Thorax—The middle part of an insect to which the wings are affixed.

Ungulata—Animals that have hoofs.

Univalve—Having but one shell.

Vertebrata—Animals that are furnished with backbones.

Viviparous—Producing young from eggs while in the body, as in the viper.

Wattles—A fleshy excrescence that grows under the throat of the male farmyard fowl.

Zoophytes—Forms of life being mediums between animals and vegetables.

INSECTS, FISH, BIRDS, AND QUADRUPEDS
REFERRED TO IN THIS VOLUME.

WONDERFUL ANIMALS.

CHAPTER I.

CURIOUS ODDS AND ENDS ABOUT ANIMALS.

Though birds, beasts, and fishes of every kind,
Are lower than man in the scale of creation,
Their habits all show the existence of mind,
Of will, love, and hate, and parental devotion.

N perusing the pages of history, especially those of biography, we may find numerous proofs of the correctness of the old saying, that 'Truth is stranger than fiction.' This assertion is also corroborated by many of the actions and life of some of the members of the animal kingdom.

It is well known that many human beings indulge in very singular habits, while their conduct generally is often both irregular and strange. Without attempting to assign the particular causes of human eccentricities, we may confidently state that, much as the lower animals are under the influences of the law of instinct, there are instances in which they have so far diverged from their ordinary course as to be considered, if not in all cases eccentric in the true sense of the word, yet so distinguished by certain peculiarities of structure, habits, and social life as to deserve special attention, and as revealing to us some interesting and wonderful phases of animal life. The odds and ends to which we refer we will now introduce.

GOVERNMENT AMONG ANIMALS.—'A regularly constituted form of government, elementary as it may appear, exists among very many families of animals and insects. Reptiles are usually solitary, exhibiting neither affection nor social feelings. A blind

buffalo on a Western prairie has been known to act as an
absolute sovereign over a vast herd, controlling their movements as seemed to suit his own views of what was best for
the common good.   Horses, too, in their wild state, wherever
found, invariably obey the behests of a powerful stallion, who
parades his forces, forms lines of defence, or suddenly gives
orders for a stampede, as circumstances require.

' Dogs, left to themselves, establish an oligarchy, whether in
Asia, Africa, or any other continent.   The supreme authority
is invested in certain
individuals, managing
a prescribed territory,
and woe to those tres-
passing upon their do-
main !   On concerted
occasions they all act
together like wolves
for the accomplish-
ment of a grand design.
Wolves separate as
soon as they have
accomplished their de-

The Wolf.

sign; but, unlike dogs, seem not to recognise a particular leader
on their foraging expeditions.

' Grain-eating birds form associations.   Wild geese have an
admirably organized system of government.   Migrating
feathered races associate in autumn for common safety in
their annual flights ; but carnivorous birds, as hawks, eagles,
etc., are unsocial and selfish.   Domestic fowls divide into
families, at the head of which is a vigilant cock that watches
his charge with Argus eyes.   Ants, honey-bees, and wasps
form regular sovereignties.'

ECCENTRIC CRANE.—We have read of a male and female
crane which, having been procured when very young, were
taught to follow their owner wherever he went.   They became
masters of the farmyard.   Bulls, cows, and foals were subject
to their control, but they declined to interfere with the pigs.
When the female died, the male left the farmyard, and was
found in the neighbourhood two or three days afterwards in a
dejected condition.   When he was brought back he made the
acquaintance of the bull, which he accompanied wherever he

went, and would keep off the flies while the animal grazed. If the bull did not appear in time, the crane would fetch him.

When horses were put to the carriage, he would, ostler-like, stand before them, and by blows from his bill and outspread wings, prevent them from moving until they were ready to start. He was fond of the cook, who fed him, and would never go to bed until she took him under her arm and conveyed him to his sleeping-place. On one point, though as a rule very fierce and unforgiving, he was a bit of a coward. He never could endure the sight of any black moving object, such as a dog, a cat, or a crow, and his greatest dread was the chimney-sweeper.

The Crane.

THE KITE PUNISHED SEVERELY.—'When the brutes adopt the vices of men, it seems that they must look to men for protection from their kind. According to an Indian paper, a sergeant's qua-drille party lately left, in their half-emptied glasses, an opportunity to a kite for getting drunk, and this bird, the scavenger of the feathered world, appears to have made free with the "heel-taps" of the gallant party. In the morning the kite

Swallow-tailed Kite.

was found staggering about quite drunk; but after having been allowed the shelter of the mess-room for an hour or two, "the bird prepared for flight by hopping out through the door into the open air; but no sooner did it show itself than more than a dozen other kites pounced upon the drunkard and gave it a most unmerciful pecking, insomuch that it was glad to seek the shelter of the mess-room once more, and would not go out again either by force or persuasion." The conclusion of the story is that "one of the sergeants took the bird home, and it is now an inmate of his poultry-yard, well fed and fat, and evidently fond of its present quarters." It has lost caste by drunkenness, and is apparently assumed by its fellows to have passed into slavery.'

A SNAKE TEAM.—'Did you ever hear of anyone driving a pair of snakes? Mr. Frank Stockton, in his "Roundabout Rambles," tells us that boys and girls in France sometimes amuse themselves by getting up a snake-team. They tie strings to the tails of two common, harmless snakes, and then they drive them about, using a whip (I hope gently) to make these strange steeds keep together and go along lively.

Whip Snake.

'It is said that snakes which have been played with in this way soon begin to like their new life, and will allow the children to do what they please with them, showing all the time the most amiable disposition. There is nothing very strange in a trained snake. Toads, tortoises, spiders, and many other unpromising animals have been known to show a capacity for human companionship, and to become quite tame and friendly. In fact, there are very few animals in the world that cannot be tamed by man, if man is but kind enough and patient enough.'

AN AMUSING SPECTACLE.—'A curious scene was once witnessed in one of the main thoroughfares of Bolton. A donkey, drawing an empty cart, stumbled and fell down; and the carter being unable to make it get up, unharnessed it, and the cart was taken into a side street. Still, neither blows nor caresses had any effect on it; and after a while about half a dozen men lifted the donkey up, and carrying it to the cart, put it in the shafts. Here again it lay down, and was as obstinate as before;

and eventually it was placed inside the cart, and the carter, taking his place in the shafts, wheeled it home. The donkey was a remarkably large and fine-looking animal, and was to all appearance entirely unhurt.'

A TIGER AFRAID OF A MOUSE.—A tiger was confined in a cage at the British Residency in Calcutta. To poke him with a stick, or to tantalize him with shins of beef, did not annoy him half so much as putting a mouse into his cage.

This mouse, tied by a string to the end of a pole, would be pushed under the tiger's nose, when the great animal—the terror of nearly all other animals—would leap· to the other side of the cage, or jam himself up in a corner of it, where he would tremble like an aspen leaf, and roar in an ecstasy of fear. When the mouse was placed in the middle of the floor of the cage, and the tiger was forced to cross it, he would, instead of walking, leap over the mouse, and that so high, that his back would nearly touch the top of the cage.

WALKING LEAF.—This curious insect, of which · there are several species, is found in some parts of South America and the East Indies. In shape, colour, texture of the wings, the limbs spread out like small twigs bear-ing unfolding buds, the whole appearance so closely resembles a leaf,

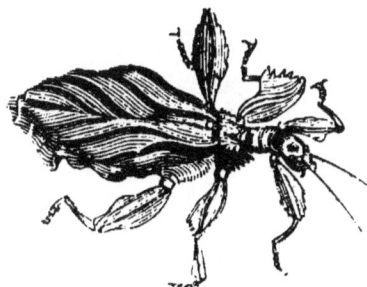

Walking Leaf.

that when hanging to a tree or bush they are not easily distinguished.

WONDERFULLY ODD.—The following information respect-ing the eccentricities of different animals has been taken from the *Evening Standard :*

'It is well known that the females of many varieties of animals have a mania for nursing the young of other species, and a correspondent of a sporting contemporary records that a Mr. Alex. Dale, of Abdie, near Newburgh, on the banks of the Tay, in Fifeshire, is in possession of a dog which is nursing a pig. Mr. Dale has a sow that had a litter of pigs ; one was taken to the house, and at once the dog adopted the little squeaker, and is rearing it with all the care, tenderness, and affection of a mother. She will not allow any stranger or animal to approach piggy. "About a year ago," the writer continues, " the same dog nursed a kitten. I have known of a dog nursing a rabbit, and of cats nursing rabbits, and one case has come under my notice of a goat nursing a calf." Hens will frequently hatch ducks' eggs, and experience the greatest anguish when their adopted children take to the water and display an accomplishment of which their foster brothers and sisters are devoid. Perhaps the mania for adopting the young of other animals is nowhere so marked as in the human race. The males and females of this species almost habitually adopt and provide for the young of dogs, cats, birds, and make pets of young horses, lambs, goats—even in one or two known cases of spiders ; while, on the contrary, two ancient Romans—if those who built Rome can be so termed with propriety—were cared for and partially educated by a wolf.'

Dog and Portrait.—The following curious anecdote, given by E. T. Evans, appears in *Science Gossip :* ' Some years ago we had a Pomeranian (dog) who took a particular dislike to a portrait of my grandfather, which hangs in the dining-room ; sometimes she would jump up and bark at it without any apparent reason, but if the wind made a noise in the chimney, she would often jump on to the sideboard (over which the portrait hangs) to get at it. Any noise whatever that she did not understand, she used to refer to this picture, and bark accordingly. I may mention that the eyes in the portrait are very well done, and seem to look at you wherever you stand ; 'this may have something to do with it.'

A Philosopher on Birds and their Oddities.—' A philosopher, who has noted the wearisomely monotonous proceedings of the young of the human race about that period of their existence when they begin to suffer from love's young dream, has turned his attention to the love-making of the young of other species. The results of his investigations he is good enough to give to the world at considerable length, in the columns of a foreign journal, and they are decidedly interesting. Birds and beasts seem to be more constantly sincere and less selfish than humanity, though it cannot be denied that some creatures—the North American grouse and heron, for example—are vain, and at times ridiculous. At

certain periods of the year the grouse meet together at a given
spot, and go through a variety of performances that show their
activity and grace. They "run round a ring, now to the
right and now to the left, jumping into the air and then hopping
on one leg." Of the herons, Audubon records that "they stalk
up and down before the females"—much as well-dressed men
appear in the Park—"showing themselves off, and bidding
defiance to all rivals. In the midst of a dignified walk they
will stop and caress some particular female, and the next
moment will be knocked over by a larger rival who does not
intend to be 'cut out;'" and he certainly stands a better
chance with the "fair one," as the females prefer handsome
consorts. The wisest and most thoughtful of all appears to
be the satin-bower bird, which builds a beautiful residence,
and decorates it carefully and luxuriously. Into this he
invites the lady bird of his choice, and shows her what the
naturalist declares are "evident instances of design." Touched
by the consideration evinced on her behalf, she usually
consents to take up her residence with the accomplished
builder. "The courtship of the great English bustard is an
extremely interesting sight," we are told. "The love-making is
done entirely in the air. Now the male will sail around in

curves, dart up, and
hover over the female,
then drop almost to
the ground, only to
rise again and con-
tinue its odd and fan-
tastic 'love-making.'
Similar in its actions
is the *Otis bengalensis*,
an allied bustard. At
such periods he rises
perpendicularly into
the air with a hurried
flapping of his wings,
raising his crest, and
puffing out the

Bird of Paradise.

feathers of neck and breast, and then drops to the ground.
He repeats this manœuvre several times successively, at the
same time humming in a peculiar tone. Such females as

happen to be near obey his enticing summons, and when they approach, he trails his wings and spreads his tail like a turkey.'" Birds of Paradise are extremely human in some of their proceedings. The females always try to select the most beautifully-coloured male birds. When in their best plumage, long delicate feathers surround these creatures like a golden halo, in the centre of which the bright green head forms an emerald disc, and when the birds moult and lose their feathers—just as, sometimes, when husbands lose their fortunes—the wives desert them. The story of the devoted attention paid to the infirm wife by a Guinea sparrow is a noble example for humanity.'

MEDICINE AS PRACTISED BY ANIMALS.—In the *British Medical Journal* we read the following remarks on how animals doctor themselves in sickness:

'M. G. Delaunay, in a recent communication to the Biological Society, observed that medicine, as practised by animals, is thoroughly empirical, but that the same may be said of that practised by inferior human races, or, in other words, by the majority of the human species. Animals instinctively choose such food as is best suited to them. M. Delaunay maintains that the human race also shows this instinct, and blames medical men for not paying sufficient respect to the likes and dislikes of the patients, which he believes to be a guide that may be depended on. Women are more often hungry than men, and they do not like the same kinds of food; nevertheless, in asylums for aged poor, men and women are put on precisely the same regimen. Infants scarcely weaned are given a diet suitable to adults, meat and wine, which they dislike and which disagrees with them. M. Delaunay investigated this question in the different asylums of Paris, and ascertained that children do not like meat before they are about five years old. People who like salt, vinegar, etc., ought to be allowed to satisfy their tastes. Lorain taught that, with regard to food, people's likings are the best guide. A large number of animals wash themselves and bathe, as elephants, stags, birds, and ants. . . . In fact, man may take a lesson in hygiene from the lower animals. Animals get rid of their parasites by using dust, mud, clay, etc. Those suffering from fever restrict their diet, keep quiet, seek darkness and airy places, drink water, and sometimes even plunge into it. When a dog has lost its appetite it eats that species of grass known as dog's grass (*chien-dent*), which acts as an emetic and purgative. Cats also eat grass. Sheep and cows, when ill, seek out certain herbs. When dogs are constipated they eat fatty substances, such as oil and butter, with avidity, until they are purged. The same thing is observed in horses. An animal suffering from chronic rheumatism always keeps as far as possible in the sun. The warrior ants have regularly organized ambulances. Latreille cut the antennæ of an ant, and other ants came and covered the wounded part with a transparent fluid secreted from their mouths. If a chimpanzee be wounded, it stops the bleeding by placing its hand on the wound, or dressing it with leaves and grass. When an animal has a wounded leg or arm hanging on, it completes the amputation by means of its teeth. A dog, on being stung in the muzzle by a viper, was observed to plunge its head repeatedly for several days into running water. This animal eventually recovered. A sporting dog was run over by a carriage. During three weeks in winter it remained lying in a brook, where its food was taken to it; the animal recovered. A terrier dog hurt its right eye; it remained lying under a counter, avoiding light and heat, although habitually it kept close to the fire. It adopted a general treatment, rest and abstinence from food. The local treatment consisted in licking the upper surface of the paw, which it applied to the wounded eye, again licking the paw when it became dry. Cats also, when hurt, treat themselves by this simple method of continuous irrigation. M. Delaunay cites the case of a cat which remained for some time lying on the bank of a river; also that of another cat which had the singular fortitude to remain for forty-eight hours under a jet of cold water. Animals suffering from traumatic fever treat themselves by the continued application of cold, which M. Delaunay considers to be more certain than any of the other methods. In view of these interesting facts, we are, he thinks, forced to admit that hygiene and therapeutics, as practised by animals, may, in the interests of psychology, be studied with advantage. He could go even further, and say that veterinary medicine, and perhaps human medicine, could gather from them some useful indications, precisely because they are prompted by instinct, which are efficacious in the preservation or the restoration of health.'

CURIOUS COMPARISONS.—How often the names of animals

are used to express opinions entertained of the virtues, vices, habits, and dispositions of men. For instance, one man is said to be 'as stupid as an *ass*,' another, 'as busy as a *bee*,' or 'as savage as a *bear*,' 'as nimble as a *cat*,' 'as harmless as

a *dove*,' 'as slippery as an *eel*,' 'as cunning as a *fox*,' 'as green as a *gosling*,' 'as sharp as a *hawk*,' 'as mad as a March *hare*,' 'as bold as a *lion*,' 'as poor as a church *mouse*,' 'as foolish as a *moth*,' 'as strong as an *ox*,' 'as plump as a

Eel.

partridge,' 'as fat as a *porpoise*,' 'as dirty as a *pig*,' 'as proud as a *peacock*,' to be '*pigeon*-toed,' 'as weak as a *rat*,' 'as silly as a *sheep*,' 'as fierce as a *tiger*,' 'as full as a *tick*,' 'as ugly as a *toad*,' and 'as keen as a *wasp*.'

A man who gets drunk is called 'a drunken *dog*,' and said to be 'as dry as a *herring*.' A man with a bad temper is called 'a snarling *dog*;' one with a good temper, 'a jolly *dog*.' A discontented man is 'a grumbling *dog*;' those who are lacking in good moral principles are said to be 'bad *dogs*;' the man who will not be convinced of error is considered to be '*pig*-headed,' or 'as blind as a *bat*.' Ladies who talk fluently are said to 'chatter like *magpies*,' and the tongues of men who are very loquacious are said to

'wag like *lambs'* tails.'

CURIOUS FACTS. — The *Building News* observes that 'bees are geometricians. The cells are so constructed as, with the least quantity of material, to have the largest-sized spaces and the least possible loss of interstice.

Beaver.

The mole is a meteorologist. The bird called a nine-killer is an arithmetician; as also the crow, the wild turkey, and some other birds. The torpedo, the ray, and the electric eel are electricians. The nautilus is a navigator. He raises and lowers his sails, casts and weighs anchor, and performs other nautical acts. Whole tribes of birds are musicians. The beaver is an architect, builder, and wood-cutter. He cuts down trees, and erects houses and dams. The marmot is a civil engineer. He not only builds houses, but constructs aqueducts to drain and keep them dry. The white ants maintain a regular army of soldiers. Wasps are paper manufacturers. Caterpillars are silk spinners. The squirrel is a ferryman. With a chip or piece of bark for a boat, and his tail for a sail, he crosses a stream. Dogs, wolves, jackals, and many others are hunters. The white bear and the heron are fishermen. The ants have regular day labourers.'

Jackal.

CURIOUS COINCIDENCES.—We have often noticed the appropriateness of the names of some commercial men to the trades followed by them. For instance :

There's *Steer*, a well-known butcher ;
And *Bull*, a large cheese factor ;
There's Mr. *Pike*, the fishmonger ;
And *Hare*, the game contractor.

There's Mr. *Duck*, the poulterer ;
And *Hide*, by trade a skinner ;
A Mr. *Lamb*, who deals in wool ;
And *Nightingale*, the singer.

There's *Fish*, the herring-curer ;
One *Roebuck* makes horn handles ;
A Mr. *Whale* sells oil and fat,
Of which we make our candles.

There's Mr. *Fox*, the furrier ;
One *Jay* sells ostrich feathers ;
Swan deals in costly eider-down ;
Rabbits, in coloured leathers.

A Mr. *Gosling* sells quill pens ;
One *Beaver* is a hatter ;
John *Peacock* keeps all kinds of birds
Who twitter, sing, and chatter.

CHAPTER II.

PEEPS DOWN A MICROSCOPE.

Some animals are far too small
    For any human eye to see
Unaided by a microscope :
    Then come and peep down one with me.

' All are but parts of one stupendous whole,
Whose body Nature is, and God the soul.'

POPE.

HERE is much that is wonderful in the organization, not only of the largest, fleetest, strongest, and most beautiful of our animals, but in that of the tiniest form of life. The difficulty in manufacturing any kind of machinery is, as a rule, in proportion to its smallness. Great care, delicacy of touch, and exactness, as well as marvellous mechanical genius, are more necessary in making a minute piece of mechanism than in that of a large and ponderous one. In the smallest members of the insect world we may see as great a distinction between the different parts of their microscopic bodies as in those of the elephant, horse, lion, or a human being. Insignificant as insects may appear to be, it should not be forgotten that what God has condescended to create is worthy of the notice even of the highest born of mortal beings, be he emperor, prince, or peer.

If in the warm summer-time we stand by the side of a pool of stagnant water, we may see moving masses of a pale, or deep-red, green, or yellow colour. These consist of numbers of

ANIMALCULÆ, which are so minute as not to be seen without the aid of a microscope. It has been ascertained that a drop of the scum, not much larger than a pin's head, taken from stagnant water, often contains one hundred separate existences, which move and frisk about with amazing rapidity, and have been seen to represent several distinct and different species, all as full of life as animals in the higher grades and of larger magnitude. They twist about with surprising activity in search of food, the larger kinds showing their voracious appetites by devouring the smaller species.

It is stated that these minute creatures possess eyes, mouths, and stomachs, as well as feet, nerves, and muscles, all covered with bristles or a tegument of some kind, which serve not only for protection, but for ornament. When we look at the large number of these creatures dancing about in so contracted a space, we must admit that the most finished specimen of man's genius and work bears no comparison with them, either in variety of form or in wonderful structure.

A MOSS-COVERED WALL, ETC.—In the following experiment we may, by the aid of the microscope, bring to light one of the most wonderful and mysterious things in nature. ' Take a handful of dry moss from an old wall, as dry as you can get it. Moisten it with distilled water, then squeeze it; the drops at first will be a little thick . . . let these drops lie on a piece of glass ; at first there may be no sign of life. In a few minutes squeeze the moss again. Small yellow spots of an irregular oval form will then appear on the glass. You will then see these forms gradually lengthening, bulging out at each end, and assuming the shape of a caterpillar, only that one end will be more tapering than the other. Afterwards this end will send out a fork which becomes firmly attached to the glass, while the whole body sways from side to side. The head will then be drawn in as if buried, when two tooth-like wheels at once make their appearance, rotating rapidly ; from this they are called "Rotifera." Thus there is a resuscitation from death, not from drowning, but from dryness.' It may be said that this speck of animated matter has been drowned into life.

The crowds of infusoria revealed by the microscope are, in their forms, of the most marvellous kind, and may be seen gliding past or sporting in a mazy dance ; but ever and anon there comes rushing among their swarms, like a fierce tiger through a flock of sheep, some monster of a different kind, having on its head what appear to be great wheels, that continually spin round and round, and, like the paddles of a steamboat, serve to move it through the water. The animals in question, as before mentioned, are named 'rotifera,' or 'wheel-bearers.' In their size they much exceed the humbler infusoria, over which they tyrannize. Their length may be roughly estimated at about one-fiftieth to one-hundredth of an inch—terrific giants when compared with the small fry around them, although themselves scarcely perceptible by unassisted vision.

LIFE IN OLD SPOUTS.—If we take a little of the red earth found, in dry weather, in old spouts, or dried-up drains, we may not perceive any sign of life in it. Even if we spread a handful of this red earth upon a smooth surface, however carefully we may look at it, nothing like moving life presents itself; it has, in all respects, the appearance of nothing more than cold inanimate earth or sand, the grains of which are so light and dry that a breath of wind may scatter them in all directions.

Let this earth, however, be gathered up, and well moistened with water. It will be seen, in the course of a little time, that it has undergone a marvellous change. It now contains forms of active life, so small that the keenest human vision cannot, unassisted, detect them singly, a fact which renders it difficult minutely to describe them.

According to Rymer Jones, the following curiously formed living organisms were discovered, by the aid of the micro-scope, on a piece of duckweed taken from a pond. 'Some of these creatures were of a trumpet-shape, around whose gaping mouths whirled the swarming atoms they had to swallow. Others, like wine-glasses in miniature, stretched out the little bells that constituted their bodies in search of food, and, when alarmed, shrunk timidly from danger. Some had the shape of rolling mulberries, that gently made their way through the water. Others, formed like swans, glided up and down with graceful elegance. Some shot about like meteors; and many, clad in shells, and armed with leg-like hooklets, skipped from point to point like living scintillations.'

'THE EGGS OF ROTIFERA form beautiful objects for micro-scopic study. They are covered with a transparent shell, through which the parts of the embryo, as they develop them-selves, gradually become distinctly apparent, until at length the cilia are seen performing their mimic rotation, though as yet the imprisoning shell has not been broken. At last, by the action of these organs, which every moment become more energetic, the transparent membrane is ruptured, and the little creature bursts forth, eager to enter upon its new existence, and already possessing the form of its parent. The time from the exclusion of the egg to the hatching is commonly about twelve hours. Ehrenberg watched an individual through eighteen successive days; it was full-grown when he first observed it, and it did not die of old age at last. Such an individual he found to be capable of producing four eggs every twenty-four hours, the progeny derived from which grow to maturity and exclude their fertile ova in the same period; a single rotifer thus producing in ten days forty eggs, developed with the rapidity just stated; this rate, raised to the tenth power, gives one million of individuals derived from one parent, on the eleventh day four millions, on the twelfth day sixteen millions, and so on. Well may our ponds and ditches swarm with their multitudes, and countless creatures dependent on such a supply rejoice at the abundance of food thus supplied to them!

'All the rotifers have a marvellous fund of vitality, and survive under circumstances where animals less tenacious of life would die a thousand deaths. They have been thoroughly dried by means of chemical acid, wetted, and restored to life, dried again, wetted again, and subjected to this treatment through many successive alternations without perishing.'

'Microscopical investigation,' says Rhind, 'is continually adding fresh wonders to our knowledge of these interesting atoms (known as *Infusory Animalcules*), and furnishing fresh proofs of the amazing power and wisdom of Him who made them all. These minute creatures are various indeed in their shapes, structure, and habits; some inhabit fresh water, some the sea, some frequent the surface, some revel in the lowest depths of the ocean. Many are protected by delicate shells, others are otherwise cared for; some inhabit the fluids of animals, some are found in the cells of plants; some are provided with organs of motion, others remain attached to fixed or floating objects. The propagation of some is by eggs; in others by division of the parent; and in some by sprouting buds. There can be little doubt that the air is always carrying about numbers of the germs of animalcules; and that, when they fall into water in a state suitable for their development, they vivify and reproduce.'

Our poet Thomson, in referring to this countless family, says:

' Whence the pool
Stands mantled o'er with green invisible
Amid the floating verdure millions stray.'

How can we look upon the structure of these minute beings without being humbled and lost in admiration of their Creator, who sustains and notices them as well as He does worlds of suns, and systems of stars and planets ?

ANIMALCULÆ IN EVERYTHING.—But microscopic forms of life not only invade earth, air, and water, but are found in the interior of animals and plants. Even man cannot escape them ; his mouth contains them in the tartar that loosens his teeth. Legions of worms, imperceptible without scientific aid, are found in our fleshy structure ; and as many as twenty-five of them have been counted in one of the muscles of the ear which does not exceed a grain of millet in size.

Worms not larger than a pin's head accumulate in the head of the sheep, causing staggers and ultimately death.

It has been admitted that man's ingenuity, all his inventions, his precautions, his medicines, and scientific appliances have failed to exterminate this microscopic life. Its dominion has no bounds, ' it is immensity itself.'

CHAPTER III.

LILLIPUTIAN SUBJECTS OF THE ANIMAL KINGDOM.

Though genius may deserve our praise,
And art may sculptured statues raise
    Of animals and man ;
Insects are greater far than they :
Each moves, hears, sees, and has its day
    Of life, though short its span.

NO one who has a contemplative mind can look on Nature without being impressed with the grandeur, majesty, and beauty of all her works, and convinced that infinite power has created them, that they are under the control of perfect laws, which have emanated from infinite wisdom. According to the character or magnitude of the object on which the eye may gaze, so will be the impression and effect produced upon the mind.

Who can look upon the restless ocean, and not say with the poet,

' Beautiful, sublime, and glorious !
    Wild, majestic, foaming free !
Over time itself victorious,
    Image of eternity !'

or upon gigantic rocks, and not be reminded of the immutability of the Creator ? Even the fruit of our fields and orchards, blushing flowers, and sweet feathered singers will help to make the heart respond in gratitude to their generous Giver.

There are, however, in these things, as well as in quadruped and bird life, much hidden beauty, which the intellect of man is too dull either to recognise, comprehend, or explain ; but in nothing is this more prominently shown than in the existence of the complex, multifarious, but minute creatures which constitute the subjects of the insect world. Here is enough to fill the deepest thinker, the most learned philosopher, and the most ardent lover of nature with amazement.

We shall now make a few general remarks on these Lilliputian forms of life, and then refer to the orders in which they are arranged, also to their beauty, and a few other interesting particulars respecting them.

WHAT IS AN INSECT?—Insects belong to the third class of articulated animals, whose bodies are divided into three distinct or principal portions, known as the *head*, which contains the senses, the mouth, and antennæ. The *thorax*, or middle portion of the body, on which are six, and sometimes four, legs and two wings. The *abdomen*, usually the largest part of the body, contains the viscera, connected with nutrition and reproduction, but is without legs.

In the 'Treasury of Natural History,' we are informed that 'Insects surpass in variety of structure and singularity of appearance all the larger branches of the animal kingdom. The general characters by which they are distinguished from other animals are these : First, they are furnished with several feet ; secondly, the muscles are affixed to the internal surface of the skin, which, though hard, sometimes preserves a certain degree of flexibility ; thirdly, they breathe, not like the generality of larger animals, by lungs or gills, but by spiracles or breathing-holes, distributed in a series or row on each side the whole length of the abdomen, and communicating with two long air-pipes within their bodies, and a number of smaller ones to carry the air to every part.

'Insects have a very small brain, and instead of a spinal marrow a kind of knotted cord, extending from the brain to the hinder extremity ; and numerous small whitish threads, which are the nerves, spread from the brain and knots, in various directions. The heart is a long tube, lying under the skin of ·the back, having little holes on each side for the admission of the juices of the body, which are prevented from escaping again by valves or clappers, formed to close the holes within.'

Although the ancients entertained an idea that insects were bloodless, this has been proved to be erroneous. 'It is said that in the hearts of insects there are several chambers, divided by transverse partitions, in each of which there is a hole shut by a valve, which allows the blood to flow only from the hinder to the fore part of the heart, and prevents it from passing in the contrary direction.

'The blood of insects differs from that of the larger animals chiefly in colour, since in most insects it wants redness, being generally of a clear and watery aspect, and sometimes of a yellowish hue.'

Newman says: 'The senses of insects are seven—love, touch, taste, smell, hearing, sight, and the commanding and governing sense called volition, mind, thought, or instinct,' to the last of which we will briefly call attention.

INSTINCT OF INSECTS.—'What is instinct?' was a question we put to a friend a short time ago. The answer was, '*Reason stereotyped.*' As this may not be sufficiently definite to some of our readers, we offer the following remarks, which may help to elucidate the answer referred to.

Instinct is a psychical property with which animals are endowed, that prompts them to perform certain acts under the guidance of their senses, such acts tending to the well-being of the individual and maintenance of the species. Many instinctive actions are performed entirely without education or experience, and some suppose without a knowledge of the end to be attained. We think, however, the last sentence requires considerable qualification, and that while admitting it may be true in many cases, there are others in which it is the contrary.

In the infantine stage of animal existence, when reason is nil, and no knowledge has been gained by experience, the impelling power of instinct may be seen in full force. A new-born child adopts the right means to obtain from the mother the sustenance it needs, but it does so without a knowledge that it is to support its life, and to build up its structure. The same remarks apply to the young of mammalia, and of all other living creatures.

As animals reach maturity, instinct, though not destroyed, appears to give place to developing reason; and experience becomes the prompter to many of the actions and habits of the lower creatures, which they seem to know will produce the results they desire to see.

Even admitting the white and tortoiseshell butterflies, which deposit their eggs upon the cabbage and nettle, may have no knowledge that those plants constitute the proper food for their offspring, we can hardly suppose that birds build their nests, or that foxes, badgers, moles, and other underground animals make burrows in the earth, without knowing that these are a necessity of their nature, and intended for their own comfort, safety, and protection, and also for the benefit of their future young.

It may be true that, unlike man, who learns and improves by experience, these creatures may not do so to such an extent either in their architecture or mining operations; yet it is no valid argument against our theory that the animals referred to know that they are working, in mining and building, for a certain purpose.

It may be difficult to prove that these remarks apply with equal force to *insects,* and yet we are not without reason for supposing that even they know it is necessary under all circumstances to use those means placed in their power, and to have recourse to various stratagems to defend themselves, and to find support for their progeny. We have a proof of this in the

BOMBARDIER BEETLE.—It appears that these insects are able to alarm their enemies by means of real artillery. 'These Coleoptera, when threatened, suddenly expel from their intestines a whitish acid vapour, the explosion of which, as it issues, produces a certain sound, a slight detonation, which carries disorder among the aggressors. Hence, when one of these insects is pursued by an enemy, it fires off its artillery anew. At the sound of a cannon-shot from one of them, all the others fire at the same time; there is a running fire along the whole line.'

Rove Beetle.

As these beetles fire in the way described for self-defence, and in order to repel their pursuers, who are other beetles of a larger kind, they must have some notion, not only of danger, but of the necessity of using the proper means to accomplish these objects, and must also be conscious of the end to be attained by so doing, or why fire at all?

THE SPIDER must surely know the object for which it spins

its web, or why should it hide in ambush until some unfortunate fly is caught in the meshes this little weaver has so
ingeniously woven? It waits patiently in expectation of a
meal, to be obtained by the plan it has adopted. To do one
thing to gain another seems to us to imply a knowledge of
both. When the spider discovers a part of its web is broken,
it at once proceeds to repair it, as if conscious that neglect
to do so would lessen its chances of ensnaring its victims.
The following fact is, we think, strongly in favour of our
theory. When

AN ANT'S NEST is disturbed, the occupants show the
greatest concern for the safety of their eggs, which they carry
down into the lower chambers of their dwelling. But why do
they do this if they have not an idea that their eggs will be
safer down below than if left exposed on the surface? They
seem to know that this is the best plan to save their eggs, and
therefore wisely adopt it.

MISTAKES OF INSTINCT.—Whilst it is true that instinct
guides an animal to defend itself against its enemies, to use
the right means to obtain food, and to select that which is
the most suitable for its sustenance, there are no doubt
instances of what we may term *mistakes of instinct*, occurring
when it depends purely upon the information of the senses,
especially in those cases when two particular objects, strongly
resembling each other either in size, form, or colour, are presented before the animal.

It is said that 'some animals are deceived by artificial
resemblances; the trout and salmon rise at an artificial fly,
and the pike snaps greedily at the glittering metallic bait.
The edible frog, which rejects dead food, seizes a dead mouse
if it be moved gently to give it the appearance of life. Many
animals mistake the representation of an object for the object
itself: the drawing for the original. A tame roller (the bird
Coracias Abyssinica), at large in a room, pecked at the drawing of a locust for the actual insect. . . .

'In the account of that singular parasitic flower, the Rafflesia
Arnoldi, it is mentioned that it smells like tainted beef, and
that swarms of flies have laid their eggs upon it.' When the
fire-flies in Jamaica settle upon the ground, the bull-frog
devours them; to these frogs red-hot pieces of charcoal have
been thrown at dusk, which have been immediately swallowed,

causing death, the frogs mistaking them for fire-flies. But human beings have been deceived in like manner. The nutty flavour of cherry-laurel water and of prussic acid are so much alike, that inexperienced people have been known to take the latter for the former, and have died in consequence.

WHAT KIRBY SAYS.—This writer observes, ' That the instincts of a considerable number of insects are endowed with an exquisiteness to which the higher animals can lay no claim. What bird or fish, for example, catches its prey by means of nets as artfully woven and as admirably adapted to their purpose as any that ever fisherman or fowler fabricated ? Yet such nets are constructed by the race of spiders. What beast of prey thinks of digging a pitfall in the track of the animals which serve it for food, and at the bottom of which it conceals itself, patiently waiting until some unhappy victim is precipitated down the sides of its cavern ? Yet this is done by the ant-lion and another insect; and even the dwellings of the beaver, and the hanging nest of the tailor-bird, exhibit less wonderful and elaborate indications of instinct than a society of bees, with all their peculiar arrangements for the future, and the good order of a numerous society of different gradations of rank and utility.'

When about to spin their cocoons, insects select suitable places for that purpose ; but if prevented in carrying out their usual plan by being placed in unfavourable circumstances, they will use the best means that may be in their power. A caterpillar was once isolated in a basket, the bottom of which was not suitable for the insect to rest upon, and therefore it refused to spin. But a leaf having been thrown in, it was at once seized, then rolled into a hollow form, and otherwise so disposed of by the caterpillar as to induce it to spin its cocoon. This seems like a modification of instinct.

INSECTS GAIN KNOWLEDGE BY EXPERIENCE.—In corroboration of this assertion, Kirby relates an instance on the authority of Mr. Wailes, who says, ' He observed that all the bees, on their first visit to the blossoms of a passion-flower on the walls of his house, were for a considerable time puzzled by the numerous overwrapping rays of the nectary, and only after many trials, sometimes lasting two or three minutes, succeeded in finding the shortest way to the honey at the bottom of the calyx ; but experience having taught them this knowledge, they afterwards constantly proceeded at once to the most direct mode of obtaining the honey, so that he could always distinguish bees that had been old visitors of the flowers from new ones, the last being invariably at first long at a loss, while the former flew at once to their object.'

Addison, in speaking of instinct, remarks, 'I look upon instinct as upon the principle of gravitation in bodies, which is not to be explained by any known qualities inherent in the bodies themselves, nor from any laws of mechanism ; but, according to the best notions of the greatest philosophers, is an immediate impression from the first mover, and the Divine energy acting in the creatures ;

> ' " For reason raise o'er instinct as you can,
> In this 'tis God directs, in that 'tis man." '

ORDERS OF INSECTS.—Although the different orders of insects, as arranged by naturalists, are here referred to in alphabetical order, we shall mention only the names of the various kinds of insects under the appellation of the order to which they respectively belong. Even this rudimentary information is not only necessary, but an important step towards attaining a knowledge of insect life. Indeed, we may venture to say it is the foundation of real success in the study of this vast but interesting subject.

To students of natural history, and especially to those who wish to have, in detail, information respecting the Lilliputian subjects of the animal kingdom, we recommend Rymer Jones's volume, the 'Animal Creation,' and Louis Figuier's 'Insect World.'

The orders of insects are as follows :

1. *Aptera.*— Which includes not only the common *flea*, but those of every species, of which there are many. Also the

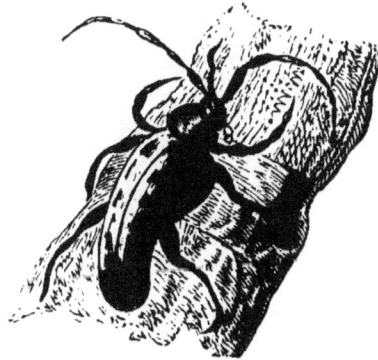

Tumble-dung Beetle.

chigoe, and *lice* of various kinds. These insects are wingless.

2. *Coleoptera.*—This order comprises a very numerous family of what are termed shield-winged insects, including the *beetle* tribes, such as *tiger-beetles, ground-beetles, cockchafers, bombardier-beetles, rove-beetles, spring-beetles, water-beetles, tumble-dung beetles, carrion-beetles, sexton* or *burying-beetles scavenger-beetles, blister-beetles, glow-worm, death-watch,* and many others. It is stated that in this country alone there are nearly 4,000 different kinds of beetles.

3. *Diptera.*—Insects included in this order are two-winged ones, such as *gnats, mosquitoes, chameleon-flies, daddy-long-legs,* and others of the *crane-fly family, whame-flies, domestic-fly, forest-fly, gad-flies, spider-flies, wasp-flies, flesh-flies, rat-tailed larvæ,* etc.

Forest Fly.

4. *Hemiptera.* — Four-winged insects, who fly quickly, but only for a short time. The two sub-orders of Hemiptera include both land and *water bugs.* The former comprises the *red-cabbage bug,* the *raspberry-grey bug, bed-bug, fly-bug ;* and the latter, what are known as *skip-jacks,* or *water-measurers,* seen in summer-time skimming the surface of water in search of drowning or even dead flies ; we may also add the *water-scorpion,* which procures its food on the stems of submerged plants, and sometimes at the bottom of the pond.

5. *Homoptera.* — Homopterous insects, or plant-suckers, are furnished with four wings, loosely veined but very transparent. In this family of Lilliputian subjects of the animal

Cochineal Insect.

kingdom, we may reckon *tree-hoppers,* known by the noise they make ; *plant-lice* or *aphides,* called the milch kine of the *ants ;* also the *cochineal* insect, valuable for the rich colour produced from its body, etc. In the same order we have *blight insects* and *lantern-flies.*

6. *Hymenoptera.*—Although insects belonging to this order have four transparent wings, these differ much in size. The hind wings seem as if they were cut out of the front ones. To this order belong *saw-flies, cuckoo-flies, ants, wasps ; bees*—such as the *honey-bee, humble-bee, carpenter*, and other *bees*, as well as the parasitical insects with which they are sometimes infested.

7. *Lepidoptera.*—The light, elegant, beautiful members of this order are also furnished with four ample wings, some of which are of the most brilliant colours. Lepidopterous insects are divided into three classes : (1) Those that fly by day. (2) Those which are seen only during the morning or evening twilight. (3) Those that are nocturnal in their habits and general activity. This family of 'little airy miracles' comprises *butterflies* of many kinds ; *moths* of various sorts, among which are the *death's-head hawk moth*, the *pack-moth*, and *feather-moth*. To the night Lepidoptera belongs the *bombyces*, from which comes the *silkworm*.

Silkworm Moth.

8. *Neuroptera.* — Insects included in this order have wings so marvellously thin and transparent, that they are known as lace-winged insects. These are *dragon-flies, may-flies, scorpion-flies, ant-lions, stone-flies, white-ants*, and *caddis-fly.*

9. *Orthoptera.* — In this order are insects that possess cutting mandibles. They are *earwigs, cockroaches, house-cricket, grass-hopper, locusts*, and *mole-cricket*, all of which have straight wings.

Locust.

BEAUTIES OF INSECTS.—A popular writer on insects says : 'To these Nature has given her most delicate touch and highest finish of her pencil. Numbers she has armed with glittering mail which reflects a lustre like that of burnished metal; in others she lights up the dazzling radiance of polished gems. Some she has decked with what looks like liquid drops of gold and silver.'

Cricket.

'In variegation, insects exceed all other classes of animated beings; some appear as if they had painted on them imitations of the clouds, rivers with their undulating waters, while others are veined like beautiful marble, and many as if they had a robe of the finest network thrown over their bodies. Among insects there are curiosities and beauties exceeding even the wildest fictions of the most fertile imaginations.'

Mole Cricket.

Man boasts of his inventions, but what are they compared to the perfect work of insects? In architecture he cannot excel—no, nor equal—many orders of insects which from time immemorial have built their houses with compartments containing staircases, gigantic arches, domes, colonnades, and the like.

Some of them have excavated tunnels so immense, compared with their size, as to be twelve times bigger than the Thames tunnel. Who, then, will not say that these animals are wonderful in their structure and beauty; that they have a claim upon our notice, and that in them there is much to excite our surprise and admiration, as exhibiting, in a marvellous degree, not only the infinite power of the Creator, but His wisdom and beneficence in their creation, their existence, and sustenance?

HABITATS OF INSECTS.—Although insects exist in every part of the world, they are variously distributed, and inhabit, according to their tribes, those countries and particular localities for which their structure is adapted, and where their needs may be the most effectually supplied. We read in 'Instinct Displayed:' 'As insects are endowed with the various powers of creeping, flying, and swimming, the air, earth, and water teem with them ; and so minute and numerous are they, that scarcely any place is free from them. Trees, shrubs, leaves, and flowers are the favourite haunts of many kinds ; rocks, sands, rivers, lakes, and standing pools, of others; whilst diffe- rent tribes, being appointed to clear our globe from all offensive substances, resort to houses, dark cellars, damp pits, rotten wood, subter-

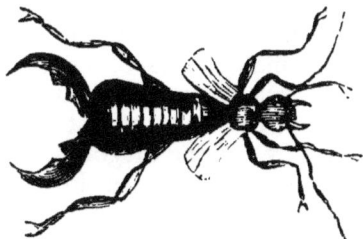

Earwig.

ranean passages, putrid carcases, and the dung of animals. These little creatures, so feeble, so diminutive, apparently so insignificant, are, nevertheless, powerful agents to benefit or injure mankind. . . . Some of them serve for food, others for medicine ; some are important in the arts, and especially to the dissecter. The great Ruysch surprised the anatomists of his day by the nicety of his preparations, which far excelled all those of his competitors. No one could imagine what means he used for this purpose, till he acknowledged that the flesh-maggot was the workman he employed, by suffering it to devour the fleshy parts.'

To the above we may add a very suggestive and humbling fact, that one of the richest materials worn even by those in the highest ranks of life is produced by a simple worm.

THE EYES OF INSECTS.—Although everything relating to insects, especially the study of their general structure, is fraught with the greatest inte- rest, it is on their eyes we wish more particularly to give some information. We are informed, that Leuwenhoeck, by the aid of a very powerful microscope,

Grasshopper.

used as a telescope, looked through the eye of a dragon-fly, and viewed the steeple of a church, which was 299 feet high, and 750 feet from the place where he stood. He could plainly see the steeple, though not apparently larger than the point of a fine needle. He also viewed a house, and could discern the front, distinguish the doors and windows, and perceive whether they were open or shut.

The eyes of insects are formed of a set of lenses, not only transparent, but so hard as not to require any covering as a protection. They appear also to be made up of a number of the smallest eyes or hemispheres, each of which, like the circles in a prismatic mirror, receives the image of the same object. The naturalist just referred to reckoned 12,544 lenses in each eye of the dragon-fly : and Mr. Hook computed that there were 14,000 lenses in the two eyes of a drone. If this be the case, then to these insects everything they look upon will appear to be multiplied thousands of times over.

How marvellous that in these minute forms of life there should exist such wonders of structure ! How poor are the most ingeniously contrived and highly finished specimens of man's work compared with the perfect works of Nature, especially as seen in the insect world!

INSECT DURATION.—Anderson says : ' The lives of some insects are in proportion to the duration of a leaf, some to that of a flower, and others to that of a plant. Earth-worms live three years; crickets, ten ; bees, seven ; scorpions, from seven to twelve ; and toads have been known to arrive even at thirty. Wasps and spiders, on the other hand, live but one year; an ephemeron, in a flying state, only one day. But naturalists speak incorrectly when, on the authority of Cicero and Aristotle, they say, that those which die at nine in the morning expire in their youth ; those at noon, in their manhood ; and those at sunset, in their old age. For, previous to their winged state, they had existed for two, if not for three years. The flying state is merely a transition which Nature has decreed to them for the greater facility of ensuring a succession.'

STRENGTH OF INSECTS.—According to their size, insects appear to be much stronger than other animals. It is said: ' If you take a common chafer or dung-beetle in your hand, it will make its way in spite of the pressure placed upon it. Accounts have been given of the very great weights that even a flea will easily move, as if a single man should draw a waggon with forty or fifty hundredweight of hay.'

CHAPTER IV.

INSECT ARMIES, AND HOW RECRUITED.

Insects flying, running, leaping,
And some slowly onward creeping ;
Everywhere we see them rife :
In themselves a world of life :
Claiming as their rightful dowers
Corn and fruit, and sweetest flowers.

IMAGINATION, in the absence of ability to look personally upon realities, may give at least a certain amount of pleasure to our minds. The power to see with the eye of the mind is, or should be, always a great boon to man, and especially a solace to those whose happiness may be marred by isolation from beloved friends.

Availing ourselves of the advantages of imagination, we will now by its aid take a stroll through our gardens, orchards, and fields ; and while the pure fresh breeze is blowing over mountain, vale, and stream, and we are inhaling the perfume of the honeysuckle, wild rose, and other flowers, and looking upon our orchards of blushing fruit, and on waving crops of golden grain, let us not be unmindful of those tiny animals which in great numbers everywhere abound, enjoying their span of life in the warm sunshine, floating in its beams, burrowing in the ground, or feeding upon the rich produce of our gardens and fields.

Having in the previous chapter referred to the different orders of insects, their wonderful structure, beauty, and ingenuity, we will now particularize some of those which are in our country remarkable for their numbers, their beautiful and

4

curious organism, and the work in which, for good or evil,
they are actively engaged.   The first we shall refer to is

THE WOOLLY APHIS.—This singular insect often abounds
in our orchards.   Although it may be at that season of the
year when nature has thrown over our apple, pear, and other
fruit trees a budding mantle of the purest and most delicate
white, fringed, perhaps, with blushing pink, we shall find on
the branches of some of these trees numerous excrescences resembling tufts of white wool, but which envelop diminutive
insects called the *Woolly Aphis*, whose stings are so severe
that trees have been killed by them, and of course rendered
useless.

'The apple-tree aphis fixes itself on the lower part of the
trunk of that tree, whence it propagates itself downwards as
far as the roots, underneath the graftings, etc.   It also likes
to lodge in cracks of the trunk and large branches.'

Such are the weapons of destruction with which some very
small insects infesting our forests are furnished, that they have
been known to open out wide clearings in them more quickly
than the axe of the woodman.

It has often happened, in an incredibly short space of time,
that gooseberry bushes have been stripped by insects of their
foliage so completely as to leave only mere skeletons, the
branches and fibrous portions or ribs of the leaves.

The name of insects more or less destructive to our fruit
and other trees is *Legion*.   Charles Muller says that the oak
supports upwards of 200 animals, which are united to it by
their parasitic existence.

Many insects live, eat, and work in our trees and forests
unseen—at any rate, unnoticed : in some cases because of
their minuteness ; in others, because their colour is the same
as the bark or leaves on which they feed ; and more frequently
because they are hidden in deep crevices of the bark, or concealed in some kind of covering, as in the case of the woolly
aphis.

A *kitchen garden* presents, during the warm summer months,
a very diversified and interesting scene of active animal life.
Flitting from herb to herb may be seen the white butterfly,
intent upon some special object or suitable place on which to
deposit her eggs.   This is a plant of the cabbage tribe, which
her amazing instinct has led her to select as the only suitable

food for her future progeny, the caterpillar. Soon after this duty has been performed the mother dies.

CATERPILLARS.—Microscopic examinations have revealed something very marvellous in the structure of the caterpillar. It is said there are only 370 fleshy muscles in man, whilst Lyonet discovered in a single caterpillar more than 4,000. Pouchet states that ' in certain caterpillars the digestive power is so great that they swallow

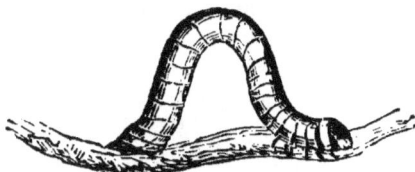

Caterpillar.

every day three or four times their own weight of food. If the elephant and rhinoceros were to feed on this scale, and were as numerous as the others, they would require only a very short time to devour all the vegetation on the globe.'

The motive parts of caterpillars consist of feet before and behind, by which they are able to go step by step, to climb up vegetables, and to reach from their boughs and stalks the food they require. For this purpose their feet are well adapted. Behind they have broad palms for adhering to the object : these have round sharp nails, by which they grasp what they want to hold by. The fore-feet are hooked, and by these they can draw leaves, etc., towards them, as well as hold the fore-part of the body while the hinder-parts are brought up thereto. These locomotive appliances are suitable and necessary only for their nymph state, as will be seen in our remarks on the butterfly.

The incredible numbers of caterpillars which have often appeared, and with little warning, within a few days in our vegetable gardens, have been not only a source of wonder, but of fear that these marauders would so denude the garden of cabbage vegetation as to leave nothing for our own consumption. So prolific are butterflies that it no doubt would be so were it not for the check put upon their increase or progeny by what are known as

ICHNEUMONS, OR VIBRATING FLIES.—Ichneumon fly is the name of a large genus of insects distinguished for their preying upon other insects, and which, as far as caterpillars are concerned, they do in the following manner : No sooner does

Nature prompt the fly to deposit its eggs than it flies off in search of a proper nidus for them. Lighting first upon one leaf and then another, walking over them and under them, she at last finds the very victim which seems to answer her purpose, namely, a butterfly or moth caterpillar. Kirby says 'that if she discovers she has been forestalled by some precursor of her own tribe . . . she leaves it, and proceeds in search of some other yet unoccupied.'

Having succeeded in finding one, she begins her work by piercing, with her sharp auger, the skin of the unfortunate caterpillar, and then inserts her eggs, probably 100 or 150 in number, underneath it. As soon as the young are hatched they begin to feed upon the fat of the caterpillar. In the course of time they attack the vital organs—the victim dies—the larvæ issue through openings and spin their cocoons on the corpse. Wrapped up in these winding-sheets of silk, they are often so numerous that the remains of the caterpillar are concealed.

THE TORTOISESHELL BUTTERFLY lays its eggs upon the leaf of the nettle, for the same reason as the white butterfly lays hers on the cabbage—namely, because the leaf of the nettle is the only suitable food for its young ones as soon as they are hatched.

THE GAD-FLY finds a future home for its offspring in the stomach of the horse. By a glutinous substance it fastens its eggs to the hair of the horse, but, we believe, on no part of the body he cannot reach with his tongue. By the warmth of the horse's body the eggs are soon hatched, and the young grubs, by their movements and desires for food, produce irritation on the skin, which causes the animal to lick with his tongue those parts thus irritated. The young grubs and even eggs are consequently taken into the mouth, and then pass into the stomach of the horse.

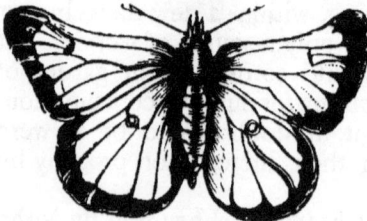

Cabbage Butterfly.

BUTTERFLIES. — Following the caterpillar through its metamorphosis to its highest or perfect state of development, we have the butterfly which in the genial sunshine delights to flit from garden to garden to gather the nectar of the sweetest flowers, and whose bright colours, transparent wings, and aërial movements have ever been the joy of infancy and childhood, and even the admiration of old age.

In the mechanism of this feeble, frail, light, and shortlived creature there is something sufficiently marvellous, not only to captivate the eye, but to astonish the mind of the most contemplative philosopher.

A German entomologist has calculated that a square inch of the wing of a peacock butterfly, as seen through a microscope, contains no less than 100,735 scales.

PAST AND FUTURE.—What a magic charm appears to be couched in the word 'butterfly!' How forcibly it reminds one of days long since past, of the frolics and gambols of boyhood, when with cap in hand we have gone in pursuit of this beautiful little fugitive, attracted by its brilliant colours and gauzy wings ; and when we were just on the point of taking it prisoner, and at that moment full of joy at the thought that we should bear away the airy prize, it has eluded our grasp and gone again in search of another flower, leaving us behind staring in utter astonishment as it sailed in the air far away out of our sight.

And how suggestive, too, is the word 'butterfly' of reviving Nature when she warms and moves into active life ! How it tells us of the departure of cold chilling winds and nipping frosts ; of spring, with its tender freshness, beauty, and plenty; of summer, with its sunshine and genial weather ; of the fruit and flowers of our orchards, fields, and gardens ; of the woods and hedgerows echoing with the music of our birds, those feathered choristers in Nature's vast cathedral ; and, in fact, of all that is bright and beautiful in this wide world ; and it even points to 'the sweet by-and-by,' when God's saints will be clothed in robes of eternal beauty, light and glory, and which will never fade away. As well as birds and bees, butterflies are in the warm summer-time the very poetry of Nature.

FOOD OF BUTTERFLIES.—As the food of the butterfly consists of the nectar of flowers, which in some cases is hidden deep down at the bottom of them, Nature has given to this insect a long proboscis, so small and light that it can easily reach the nectar without injuring, in the least degree, the flower which contains it.

The elegant and almost dazzling colours of butterflies are owing to neat and beautifully made feathers, which are placed in rows in the most systematic order. Pouchet eloquently remarks, referring to a butterfly, ' When the new creature, bursting its sepulchral laboratory, expands itself in the light, its dazzling robe reflects the brightest sheen of metal or the glitter of jewellery.'

LUMINOUS INSECTS.--There have been but few benefits conferred upon man by the aid of science greater than the discovery made of the means of producing artificial light, and especially in making it available for the many different purposes for which it is at present used. Such a boon is of great value, particularly during the long, dark, and dreary season of winter. When the light of day is declining, and the mantle of darkness is gradually spreading itself over the face of Nature, what a surprising change may be seen in our large towns and cities, which, as if by magic, are flooded, almost simultaneously, with artificial light.

For private convenience we have now lamps of many kinds —varying from the argand, the moderator, etc., to the paraffine and benzoline. These differ in mechanical construction, and in materials for burning, as well as in the amount of light they give. In fact, we have light in almost every degree, from that which is so intense that the human eye can hardly endure its brilliance, to the old-fashioned calm, quiet light of the long twelve tallow dip and the farthing rushlight.

For what we have mentioned, man is dependent on man, not only for the lights referred to, but for the lamps in which to burn them.

Nature, however, has her own lamps, which owe nothing to man for the light they give, or for the shining properties they may possess. Some of these are very wonderful, especially the moon, stars, and planets, those night gems of the sky which, as they roll onward in their respective orbits, shed their mild but useful light on our world, that without them would be little less than a dismal chaos.

Firefly.

In the insect-world Nature has also other lamps, living and

moving ones; tiny insect bodies which supply their own material, the light from which is often like the coruscations which emanate from some jewelled crown. Of luminous insects we may mention:

THE PYGOLAMPUS ITALICA.—These winged female insects are very common in Italy. Sir J. E. Smith tells us that the beaux of Italy are accustomed in an evening to adorn the heads of the ladies with these artificial diamonds by sticking them into their hair. When a number of these earth stars are seen to dart through the air on a dark night, the effect is very beautiful.

BEETLE OF THE ANTILLES.—In Cuba the women often enclose some of these phosphorescent insects in little cages of glass or wood, which they hang up in their rooms, and this living lustre throws out sufficient light to serve to work by. When travellers there find it difficult to see their way on a dark night, they will sometimes fasten one of these beetles to each of their feet.

The creoles set them in the curls of their hair, where, like resplendent jewels, they give a most fairy-like aspect to their heads. The negresses at their nocturnal dances scatter these brilliant insects over their robes of lace. In their rapid movements they seem enveloped in a robe of fire. It is the conflagration of Dejanira without the horror.

It is said that when Sir T. Cavendish and Sir R. Dudley first landed in the West Indies, and saw in the evening an infinite number of moving lights in the woods, which were merely these insects, they supposed the Spaniards were advancing upon them, and immediately betook themselves to their ships.

THE GLOW-WORM (or *Lampyris*).—Who has not seen on a summer evening these star-eyes of earth and diamonds of the night glittering in the hedgerows or under the leaves of flowering plants in our gardens, and wondered how so small a thing should throw around it such a beautiful radiance?

Glow-worm.

The glow-worm is the wingless female of a brown, softish beetle, about three-quarters of an inch long. It is something like a dark, flat caterpillar. The light, which is of a beautiful sulphur colour, proceeds from the three last rings of the body. The light seems to be without heat, and in this particular differs from sunlight or lamplight. There is, however, some difficulty in ascertaining how this luminosity is occasioned. 'This insect, during its grub state, feeds upon small slugs and such-like aliment, and is rather voracious; but as soon as it assumes its perfect form it eats only the tender leaves of plants.'

The male of this insect having wings, and the female not, a certain writer ventures to give his opinion as to the reason why; he says: 'It was necessary that some contrivance should be had recourse to for directing the rambler to his sedentary mate. What more beautiful, and at the same time sufficient guide could be possibly contrived than this self-lighted hymeneal torch?'

THE DEATH-'WATCH.'—This is a small beetle which lives on decaying wood. It is a source of great alarm and fear to superstitious people, who suppose that the ticking noise it makes by tapping its head on the wood, and which faintly resembles the tapping of the finger-nail on the table, is a sure sign or token that some near relative or dear friend will soon be taken from them by death. This noise is believed to be the ticking of death's timepiece.

THE ANT-LION.—As the description given of this insect by the writer of 'Six Days of Creation' is so clear and interesting, we will quote in full what he says: 'The ant-lion is curious for the pit formed by its larva. . . The perfect insect is very much like a dragon-fly, only with broader wings; but the larva is a wingless creature, with a pair of formidable jaws. It has little power of moving rapidly about, but it makes up abundantly for this by its surprising cunning. This larva, choosing a dry, sandy soil, constructs a pitfall with sloping sides, about two or three inches deep; and conceals itself at the bottom. No sooner does a busy ant or small beetle approach the treacherous edge than the sand slips from under its feet, and it begins to slide down the deadly slope. The falling grains of sand apprise the watching ant-lion of the success of its snare, and in a moment it jerks up a little shower of sand, the more to bewilder its hapless prey. There is little hope after once the thoughtless creature has passed the limit of firm ground; its struggles may be more or less prolonged, but its progress is ever downward, and its end the hungry jaws of its crafty destroyer.'

Pouchet, in referring to this insect and the ingenious plans it adopts to secure its prey, says : ' If the ant-lion were to keep the remains of its food near it, the snare would soon be converted into an uninhabitable charnel-house ; it must therefore get rid of them at any sacrifice. For this purpose, whenever the larva has sucked an insect, it places the corpse on its head, and then, by a vast effort, launches it into the air, and sometimes a long way from the borders of its hole, in order to obviate the suspicion which the corpses of its victims might suggest to the imprudent travellers towards the fatal refuge. In some observations which I made on the ant-lion, I saw them in this way launch flies and large ants three inches from their dwellings.'

THE CADDIS-FLY is very common about our rivers and streams, and is much used by the fly-fisher. The larva makes a curious house for its protection ; it is, in fact, a sort of tubular case made up very ingeniously of small stones, shells, fragments of the stems of water-plants, and similar things ; these it fastens together by means of a glutinous silk, which also lines the tube. Numbers of these cases may be seen at the bottom of pebbly streams, out of the ends of which the head and feet of the larva may often be seen protruding, and moving about in an irregular way.

We are informed by Rymer Jones that when full-grown this little creature creeps up the stem of some aquatic plant till the mouth of its case just reaches the surface of the water ; it then spins a net of silk across the entrance to its abode, and goes into the pupa state. At the appointed time, the insect tears its way easily through the silken grate, crawls a few inches out of the water, throws off its pupa skin, and becomes a winged caddis-fly.

THE SAW-FLY is so named because it is provided with a very curious saw in its ovipositor, with which it slits the bark of gooseberry and rose-trees, and there lays its eggs. The larvæ (often called false caterpillars), soon after they are hatched, enclose themselves in a cocoon, in which they remain some months, then change into nymphs, and in a few days into perfect saw-flies. In their caterpillar state they are so voracious, and often so numerous on the same gooseberry-bush, that they have been known to strip it of every green leaf in an incredibly short space of time. Attaining their full growth in about ten days, and being very productive, they constitute formidable enemies to the gardener, who, if it were not for the services small birds render him in devouring multitudes of these hungry creatures, would never be able effectually to keep them in check.

CRANE-FLIES constitute a numerous race, in which are included daddy-long-legs. They are found in meadows, and lay their eggs on the ground, where the grubs feed on the roots of plants.

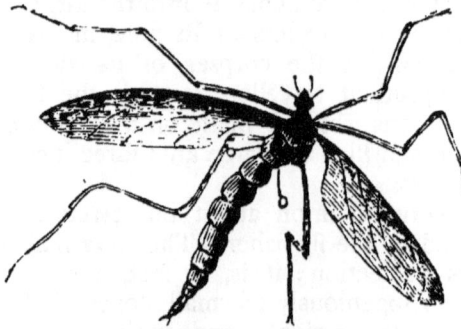

AQUATIC INSECTS. —Some of these belong to the beetle family, and may be seen almost unceasingly swimming

Tipula (Water Spider).

about in circles in our ditches and ponds. Should they, in any way, be alarmed when on or near the surface, they at once rush down to the safer depths of the water. We remember seeing, several years ago, great numbers of these aquatic insects in some water-tanks at Covent Garden Market, where we watched their active movements with great interest. The following information given by Rhind is a true description of these wonderful creatures :

' Swimming on the surface of the water, the eyes with which insects are ordinarily provided would only enable them to perceive danger from above them; but as there may arise peril from beneath, they are provided with an extra pair of eyes, two being above and two below the water, as they swim. The great diver-beetle of our ponds has two powerful oar-like legs, fringed with stiff hair. This beetle may often be seen rising slowly, tail uppermost, with its oars stretched out until it reaches the surface of the water, where it will rest. Then it may be seen that the tip of the tail repels the water, which appears to be its mode of taking breath.

Insects, like whales, have to come to the surface to breathe, or to obtain a globule of air to be consumed under the water, which is the same thing. The breathing-tubes of insects open by several apertures along the sides of their abdomen.

In the *Divers*, the horny wings fit so close as to exclude the water from coming between them and the abdomen, which would suffocate the insect; but at the end of the abdomen and wings there is a beautiful contrivance for keeping out the water while the beetle is under the surface, and for admitting air when it floats to the top. Some of these beetles are an inch and a half long, others not larger than hemp-seed, but all very voracious, feeding on other beetles, tadpoles, etc. Their state previous to assuming the beetle form is that of a formidable grub, with long legs and horrible-looking jaws, which swims about, the tyrant of the waters.

THE MITE.—Mites are a numerous race. 'Some have six legs, others eight; each leg is furnished with two small claws at the extremity, surrounded with hairs. Some of these are parasitic.'

The cheese-mite is an infinitesimally small animal often found in cheese which is of the best quality. Examined by a microscope, it will be seen to be beautifully transparent. Its eggs hatch in about twelve days, and produce, like the spider, perfect insects, or, more properly speaking, animals, which are very voracious, and change their skins several times before they become full grown.

THE FLEA.—This nocturnal disturber of the peace of man and animals lays its eggs on blankets, in the carpets of bed-rooms, or such like situations, and even

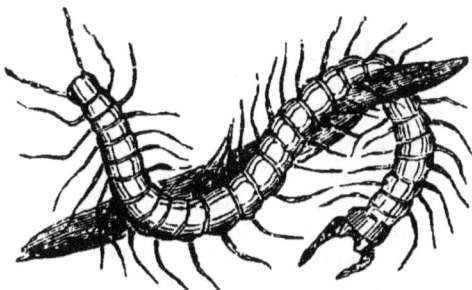

Scolopendra.

on the hairs of animals; the eggs change into maggots, then into chrysalises, and in about twenty-two days the flea is produced.

SCOLOPENDRA have bodies composed of about four-and-twenty segments, to each of which are attached a pair of active

legs.   They are furnished with curved fangs, and can instil
poison into the wound they make.   A small kind is found in
our gardens, under stones, rubbish, and lumber, and at night
sally forth in search of their prey.

THE SILKWORM (*Bombyx mori*).—As one of our valuable
articles of commerce is the produce of the silkworm, we think

Silkworm.

this insect well deserves special attention.   In the 'Animal
Creation' we find the following concise and interesting description of the worm alluded to.   The author says :

'Its caterpillar has a smooth body, and at its birth is
scarcely a line in length, but attains to even more than three
inches.   In this form the silkworm lives about thirty-four days,
and during that period changes its skin four times.   It feeds
on the leaves of the mulberry ; at the time of moulting it does
not eat, but after changing its skin its appetite is doubled.
When it is ready to change into a chrysalis, it becomes flaccid
and soft, and seeks a proper place where to make its cocoon,
in which it encloses itself ; the first day is occupied in attach-
ing, in an irregular manner, threads of silk to neighbouring
objects to support it ; on the second day it begins to multiply
these threads, so as to envelop itself ; and on the third day it
is completely enclosed in its cocoon.   This nest is formed of
a single filament of silk wrapped around the animal, and its
turns are glued together by a kind of gum.   It is estimated
that the length of the filament in an ordinary cocoon is
900 feet.   The form of the cocoon is oval, and its colour
either yellow or white.

'The bombyx remains in the chrysalis state in the interior
of its cocoon about twenty days, and when it has finished its
metamorphosis, disgorges upon its walls a peculiar liquid,
which softens it, and enables the animal to make a round hole

through which to escape. To obtain the silk produced by these animals it is, therefore, necessary to kill them before they pierce the cocoon, and then wind or reel off the thread or filament of which it is composed. To unglue it, the cocoons are soaked in warm water, then the filaments of three or four are united into one thread. That part of the cocoon which cannot be reeled off in this way is carded, and constitutes floss silk.'

THE MANUFACTURE OF SILK.—There are, we should imagine, but few persons in any part of the civilized world who do not possess some article or other made of silk. It will, therefore, be seen that an incalculable number of worms must be almost constantly employed in producing it, and that many thousands of human beings must also derive subsistence from working it up for whatever purposes it may be required.

In a work on insect architecture it is said, 'that in ancient times the manufacture of silk was confined to the East Indies and China, where the insects that produce it are indigenous. It was thence brought to Europe . . . then to Persia ; and afterwards successfully cultivated in Greece, Arabia, Spain, Italy, and France. In America the silkworm was introduced into Virginia in the time of James I., who himself composed a book of instructions on the subject.' Nearly half a century ago, it is said, a plantation of mulberry trees existed in the vicinity of London with a view to the culture of silkworms, but with little or no success. The manufacture of silk was introduced into this country in 1718, at Derby, by Mr. John Lombe.

METAMORPHOSES OF INSECTS.—Insects pass through four stages of existence, and these are generally distinctly marked. They are first contained in eggs, which are deposited by their parents in suitable situations, and with a degree of instinctive care which fills us with admiration ; they then become active and rapacious, and are well-known by the names of grubs, maggots, and caterpillars, according to the tribes to which they belong. To this condition Linnæus applied the term *larva* (which means a mask), as if the perfect insect were masked or concealed under the figure of the caterpillar. The ravages of which the forester and gardener complain result most generally from the voracity of insects in their larva state.

They eat much, increase rapidly in size, change their skins

several times, and pass into another state, in which, in some tribes, all appearance of vitality for a time is suspended. During this time some may be suspended by threads, enveloped in silk, covered with leaves or entombed in the earth, according to the habits of the species ; some of them, in this state, appear like an infant wrapped up in swaddling-clothes, hence the term *pupa* (a baby) has been given to them. After a time they reach a perfect stage of existence, and are called *imago*.

THE DUTCH PAINTER.—After looking at the pomp, the gold, and show of brilliant parties, and calling to mind the pleasure he had derived during a period of twenty years in watching the metamorphoses of insects, he has been heard to exclaim: 'Ah ! let me rather see a butterfly born.'

CHAPTER V.

AN UNDERGROUND CITY OF LITTLE PEOPLE.

Learn each small people's genius, policies :
The *ants' republic* and the realm of bees.
Mark what unvaried laws preserve each state,
Laws wise as nature, and as fix'd as fate.

NOT only are animals frequently mentioned in the Scriptures as agents appointed to carry out many of the dispensations of Providence, whether for the punishment of the wicked, or by way of rewarding the faithful and obedient, but we are invited, and even commanded, to go to them for instruction. Job says, 'But ask now the beasts, and they shall teach thee ; and the fowls of the air, and they shall tell thee ; or speak to the earth, and it shall teach thee ; and the fishes of the sea shall declare unto thee.'

Many men think it derogatory to do so, because they regard animals as being infinitely lower than themselves in mental capacity ; but the fact remains that some men by an abuse of their own faculties, their carelessness, bad habits, and indiscretion, render it indispensably necessary that they should have teachers of some kind or other, not only to instruct them in lessons of wisdom, but to reprove and to correct them for their evil propensities.

It is remarkable, and may help to teach man humility, that not only are we directed to go to the larger, nobler, and higher classes of animals to learn useful lessons, but to some of the smallest members of the insect world. The bee is presented to us as an example of untiring industry in collecting a store of honey, pollen, and propolis for future necessities. Solomon says, ' Go to the ant, thou sluggard; consider her ways, and be wise : which having no guide, overseer, or ruler, provideth her meat in the summer, and gathereth her food in the harvest.'

Let us suppose that the fertilizing showers and genial sun-shine of spring have clothed our meadows with crops of grass, now ripened and ready for cutting.   We may see in one field a group of mowers, whose scythes glisten in the sunlight, as the men with arms strong and sinewy swing them backwards and forwards, and lay the yielding swathes at their feet.

In a few hours the crop that before waved so gracefully as it was kissed by the passing breeze, is laid low, dried by the hot sun, and then carried away as a store of food for cattle in the cold dead time of winter.

We cast our eyes over the now cleared field, and our attention is arrested by numerous hillocks into many of which the scythes of the mowers have no doubt entered, and, it may be, accompanied with not 'very kind compliments' to the originators of these excrescent obstacles to the mower's pro-gress, but which are full of animal life.   If we put our spade into the base of one of these hillocks, and gently lift it up, we shall look upon a miniature town or city, compactly built, and so ingeniously arranged as to contribute to the comfort of its hundreds or thousands of inhabitants, a sight of whom may well excite our wonder and admiration.   We refer to

A NEST OR COLONY OF ANTS.—If we watch them closely we shall soon see that never did the inhabitants of any city surrounded by an invading army, and threatened with death and destruction, exhibit more excitement and concern for the safety of their lives and property than may be seen in these active and intelligent workers when thus disturbed in their underground retreat.   Their movements may seem to be all confusion, for they run one over another, some up, some down, backwards and forwards, and in every direction ; yet they are all intent upon and equally active in trying to protect their embryo young, as well as to avoid the personal danger with which they themselves are threatened.

Before we examine the structure of these members of the ant colony, or try to ascertain if in this, as well as in their doings, order, government, instinct, and habits there is any-thing to instruct or to interest us, we may notice that—

THE ANT FAMILY belong to the order Hymenoptera. There are several varieties of them, and they differ in size and colour. Some of them are of a lightish brown, and even yellow, while others are almost black. They are all, however, of similar habits, activity, and intelligence.

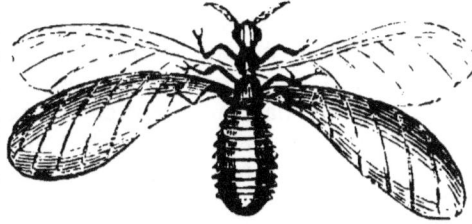

Winged Ant.

A colony of ants, like a beehive, contains three orders of occupants—males, females, and workers. The males have four wings; the females, larger than the males, have wings at the pairing season only ; the workers are without wings; indeed, at no stage of their existence have they ever been known to possess any. As there are many varieties of ants, we shall refer to some of them only. 1. *Carpenter Ants,* such as the emmet, or black carpenter ant, and the dusky ants which cut their houses out of the wood of trees. 2. *Yellow Ants,* which may be called masons, because they build their houses with solid walls, often more than a foot high, and several inches in diameter. 3. *Fallow Ants,* these are the

Fallow or Wood Ant (Female).

Fallow or Wood Ant (Neuter).

largest kind in England, and are found in great numbers in the decayed stumps of trees. Their nests are curious structures, the internal parts being composed of a heterogeneous collection of leaves, twigs, grass, and straw. The interior arrangements display considerable ingenuity and perseverance on the part

5

of these tiny builders. 4. *Turf Ants :* the mining capabilities of these ants are extraordinary. They dig chambers and galleries out of the ground from six to nine inches deep. 5. *Brown Ants :* these are the smallest of the ants, and are very remarkable for the extreme finish of their work. They are known as *Formica brunnea,* and it is to them and their habitations we shall more particularly refer.

CITIES OR NESTS OF ANTS.—Referring to the nest of the last-named ants, Kirby says : ' It is composed of earth, and consists of a great number of stories, sometimes not fewer than forty, twenty below the level of the soil, and as many above, which last, following the slope of the ant-hill, are concentric. Each story has cavities in the shape of saloons, narrower apartments, and long galleries which preserve the communication between both.'

The arched roofs are supported by thin walls, pillars, or buttresses ; some having only one entrance from above, others a second, communicating with the lower story. The main galleries in some cases meet in one large saloon, communicating with other subterranean passages which are often carried to the distance of several feet from the hill.

MATERIALS FOR BUILDING.—It is stated on good authority that these ants work chiefly after sunset, and that in building their nests they employ soft clay only, which, when sufficiently moistened by a shower, far from injuring, consolidates and strengthens their architecture. 'Having traced the plan of their structure by placing here and there the foundations of the pillars and partition walls, they add successively new portions ; and when the walls of a gallery, or apartment, which are half a line thick, are elevated about one inch in height, they join them by springing a flattish arch or roof from one side to the other. . . .

' Crowds of masons from all parts arrive with their particles of mortar, and work with a regularity, harmony, and activity which can never be admired enough. . . . They will complete a story, with all its saloons, vaulted roofs, partitions, and galleries, in seven or eight hours.' Considering the smallness of these insects, their ingenuity and industry seem more marvellous than those of bees.

Having described the structure of the ants' underground city, we shall now refer to its inhabitants, to ascertain how far they mutually contribute to the prosperity and comfort one of another. We shall notice first :

THE QUEEN ANT.—As in the case of bees, the *queen ant* is the source of life to the future colony, and it may be re-marked that soon after the pairing season is over the male ants die, so that they are never favoured or gratified with the society of their future offspring, who, it may be said, ' come into the world fatherless.' The queens lay their eggs, which are infinitesimally small, in half dozens, during which time they are usually attended by a retinue of ants, who treat them with very special respect, pay great attention to their wants, which they gratify and supply to the extent of their ability. The eggs of ants have been discovered to be of different sizes, shades, and forms. The smallest to be white, opaque, and cylindrical; the largest transparent, and slightly arched at both ends; those of a middle size semi-transparent.

THE WORKER ANTS, as their name implies, have the arduous work of the colony to perform. Their first duty, after the completion of their dwelling or nest, is to take care of the eggs laid by the queen ant. These they collect in heaps and place in separate apartments. When the larvæ appear, the workers show much thought and tenderness to-wards them. They provide them with food, remove them, when changes of the weather require it, to warmer or cooler parts of the nest, and thus render the period of their transition to a perfect state as pleasant to them as possible.

No sooner are the larvæ full-grown than they begin to weave for themselves winding-sheets or cocoons of silk, and then become chrysalides. These grubs are often so weak that they would remain hopeless prisoners in their silken gaols if the workers were not to assist them in making their escape, which they do very systematically, and with as much gentle-ness and knowledge as could be shown by members belonging to larger and higher orders of the animal kingdom.

Another duty performed by worker ants is to watch the movements of female ants. The moment one is seen in the vicinity of the nest she is at once surrounded and seized, dragged into the nest, and there retained a prisoner until she has laid her eggs. They appear also to show much concern not only in securing a queen and taking care of the embryo young, but also on the occasion of the

DEPARTURE OF MALES AND FEMALES FROM THE NEST.—
Hubert says: 'Visit a meadow on a fine summer day, and
you may see males and females of the field-ant march to and
fro over the ant-hill. They then climb all the plants near
them, but the workers always attend them. They offer them
nourishment for the last time. Disorder and excitement
increase. The winged ants climb higher and higher. The
workers follow as far as possible, running from one male to
another, touching them with their antennæ, and offering them
food. At length the whole winged tribes disappear. The
labourers return to the ant-hill, and when the weather is
favourable they make clear passages for those within that are
about to leave. They all take flight; the workers re-enter the
nest and close the entrances.'

Pairing of the male and female takes place at the outside
of their nests. They then start off to form another colony,
and to build a city for their future residence.

ANT SENTINELS.—Ants appear to pay due regard to the
safety of their underground cities, as some of them are placed
as sentinels at their entrances, and being watchful, as military
characters should be, always give notice of approaching danger.

If their nest is likely to be disturbed by the scythe of the
mower, by the upraised stick of some ignorant and mischievous clown, or by the spade of the gamekeeper, whose intention is to support the life of his game by the death of the
ants, those that are on the surface will, with astonishing
rapidity, convey the alarm to those within, and then at once
commence to carry down the larvæ and pupæ to the very
lowest apartments of their subterranean abode.

In such a republic as that we have described, and in which
so much industry, order, government, system, and intelligence
may be seen, it is natural to suppose that this underground
people possess some means of understanding each other as
to their objects, plans, and operations. This is no doubt
effected by what are called their

ANTENNÆ, OR INSTRUMENTS OF SPEECH.—The antennæ
or horns, which project from the heads of crustaceans and
many kinds of insects, serve them in the sense of feeling and
seeing. They constitute also a most wonderful and curious
part of their structure, and appear to be of essential service to
members of the ant family in various ways.

One writer informs us that: 'Previous to the military ants going out on their expeditions, they touch each other on the trunk with their antennæ and forehead. This is a signal for marching, for as soon as they are touched they are in motion. When a hungry ant wants to be fed, it moves its antennæ in a very rapid manner.

'If some moistened sugar be placed near the nest of the small black garden ant, a solitary straggler will soon accidentally discover it; he imbibes his own load, and finds his way to the nest with information; speedily, a number of others emerge, make straight for the sugar, and continue to pass to and fro in the most sedate and business-like manner, till the whole of the provender is conveyed to the nest.

'It is said that if a spadeful of earth, containing a number of ants, be thrown down in the middle of an empty room, the ants will be seen in a state of wild excitement running in all directions, and even over one another, all intent upon finding some opening in the floor by which to make their escape. No sooner is one of the ants successful in finding such a means of exit than telegraphic messages are despatched from one to another, by the lucky discoverer touching with its antennæ the first ant it meets with, and then that one touching another, and so on, until the message has been conveyed to every ant in the room, and so far understood by them that in a few minutes they all disappear down the chink or aperture so eagerly sought for.'

BRAVERY OF ANTS.—The following well-authenticated story will show that ants display a great amount of courage when they or their habitations are threatened by danger. It appears that some time ago a gentleman taking a stroll in the country where there were some ant-hills lighted his cigar with a lucifer match, which he threw, while yet burning, upon one of these hills. This soon attracted the attention of the ants, who crowded by scores round the 'blazing beam,' and then tried simultaneously to thrust it from their city. Many of them were burnt to death in their heroic endeavours to do so. Nothing daunted, the survivors pressed over the writhing bodies of their dead and dying companions, as if they were conscious that danger threatened their city so long as the burning match remained. Onward rushed the resolute firemen, who by this time had become very numerous, till at last

they rolled the match over and over and out of their pre-
cincts, charred and blackened, and incapable of further
mischief.

ANT-BEARERS AT A COCKROACH FUNERAL.—Ligon, referring to this subject, says : 'We sometimes kill a cockroach
and throw him on the ground, and mark what the ants will do
with him. Although the body is larger than many of theirs
put together, yet away they carry him, attended by other ants,
to take the place of those that may tire. Some ants act as
officers, who lead the way and show the hole into which the
cockroach must pass. If his body is crosswise and cannot
pass into the hole, it is turned endwise about a foot before
they reach the hole.' Does not this imply, not only instinct,
but the existence, in some degree, of reasoning power?

ANTS AND THE CATERPILLAR.—A writer, who takes much
interest in the habits of ants, says : 'If a small caterpillar be
placed in their way, one or two ants will at once attack it ;
but if they find they are not strong enough to master it, one
will sometimes run away into the nest and give the alarm.
Numbers of ants will come rushing out to the rescue in great
anger and excitement, which subsides the moment the prey
is slaughtered, of which the majority take no further heed,
but leave only one or two to drag the carcase homewards.'

ANTS ON THE MARCH.—The writer was sitting in his garden
one bright sunny morning when he observed a train of ants
following each other at almost equal distances, and forming a
complete line. They all appeared to come from the same
place, and to be making for the same destination, which the
author ascertained was. the case by tracking the line of ants
from the starting-point to the terminus, which was an aperture
in some loose soil in a not very distant part of the garden.

Part of the line of march extending over the surface of a
flat stone, the author crushed one of the little travellers on it
with his finger, and then drew the body about an inch across,
which left a little stain upon the stone. When the next ant
came up to the stain referred to, it suddenly stopped as if to
consider what it should do. A second and then a third ant
stopped in the same way as the first had done. After a little
deliberation they began to move, but not over the mark that
had been made by the bruised body of their comrade ; this
they refused to do, choosing rather to veer to the right hand

and to walk round the mark, keeping quite clear of it, and then turning to the left until they reached the regular line of march. For a considerable time the ants that followed took the same course. These were the common garden ants, which live under stones and make burrows in the soil.

THE WIDOWED ANT.—The self-denial and devotion of ants are worthy of special notice. We have before observed that the female ant is furnished with a pair of wings. With these she sails on the balmy air, and in the warm sunshine, with her companions, and the sight of these 'aërial dancers' is a curious and interesting one.

After the death of the male ant, the widowed bride strips off her gauzy wings, as if her airy pleasures were too trivial to be longer indulged in, and her wings too great an impediment for the very important duties she has now to perform in helping to construct her cell, laying her eggs, and cherishing her future offspring.

FOOD OF ANTS.—There has been some difficulty in ascertaining the kind of food on which ants mainly subsist. It was at one time believed they hoarded up grains of corn as provision for the winter season. So far from this being the case, no part of the ants' nest is constructed for this purpose, nor even suitable for a store of grain, nor do ants really exhibit any hoarding proclivities as they relate to corn or seeds of any kind.

It is highly probable that the larvæ found in ant-hills, having much the appearance of grains of corn, have given rise to the idea that these little colonists have sufficient forethought to furnish themselves with magazines of grain so as to enjoy the fruits of their industry through the dreary winter months.

It is, however, we believe, well-known that ants are carnivorous, feeding upon any animal food or small insects that may be accessible to them, and then conveying in their own bodies supplies of both unchanged to the labourers in the nest and other companions at home.

MILCH-KINE OF ANTS.—Ants are very partial to an insect called aphis, about one-eighth of an inch long, which extracts the juices from young and tender shoots. The lime-tree aphis ejects a viscid fluid called honey-dew, which is often found on the leaves of trees. Ants take possession of these aphides

for the purpose of getting from them a supply of this honey.
'They are in fact,' says Kirby, 'the milch-kine of the ants.'

The ants then convey these aphides to their nests, and
there furnish them with suitable green fodder, so that they
may yield all the more honey.   They guard them with great

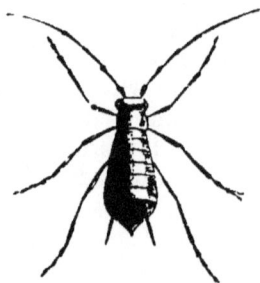

Aphis Œnotheræ (Male).   Aphis Œnotheræ (Female).

care, and will not allow the ants of another colony to visit
them, lest they should be robbed of these, their much-valued
milch-kine.

Much interesting and useful information, derived from experiments, has, during the last few years, been given by Sir
John Lubbock on ants, of which the following, taken from a
London newspaper of Nov. 3, 1882, is remarkable and curious.
It says: 'Sir John Lubbock read to the members of the
Linnæan Society an account of his further observations on
the habits of insects made during the past year.   The two
queen ants which have lived with him since 1874, and which
are therefore now not less than eight years old, are still alive,
and laid eggs last summer as usual.'

HONEY-MAKING ANTS.—The following information, given
by the *Evening Standard* newspaper of Sept. 9, 1882, may be
read with interest as being something new in reference to the
ant-family :

'An American entomologist has furnished the Scientific
Academy of Philadelphia with some rather interesting particulars concerning the habits of an insect found in certain

districts in America that goes by the name of the honey-making ant, about which very little appears to have been generally known. He started for New Mexico in pursuit of his investigations, but halted in Colorado, where, in a spot called the Garden of the Gods, he found himself in presence of a honey-ants' nest. The Garden of the Gods—so named on account of the grotesquely shaped rocks crowning the mountain-tops—is upwards of six thousand feet above the level of the sea, and covers a space of about two square miles. The sloping sides of these mountains and their summits are teeming with the honey-ants, whose homes were inspected by the American traveller. The appearance presented by the outside of the nests he describes as similar to a dyke or bank, composed of fine gravel and sand. The largest of those he visited measured thirty-two inches round its base, being about three inches and a half high. The nest was entered by a funnel-shaped passage, perforated in the centre of the dyke, and leading to a series of galleries and chambers separated into different floors. The queen ant's chamber, almost circular in form, measured four inches in diameter. The cavities of various sizes in which the ants lodge their honey are generally oval, the roofs being arched. The result of the explorer's observations in the Garden of the Gods leads him to believe that the ants are nocturnal insects, and that they gather their honey from the sweet sap of the gall-nut, storing it in the lower portion of their bodies, which resemble in shape and size a small grape. Their produce has an agreeable but slightly acid taste, and is regarded by the Mexicans and the Indians as a great delicacy. The former employ the honey also to make a spirituous drink. As the American entomologist states that it requires about nine hundred and sixty ants to furnish a pound of honey, there is no danger of their becoming rivals of the industrious bee.

ANT-WARRIORS AND SLAVE-MAKERS.—Ants are subjects of ardent passions; they are jealous of rivals, and exhibit a great thirst for booty, power, and pre-eminence, and are often engaged in quarrels and making raids one upon another. They fight with incredible fury, and display a great deal of strategy in their movements.

The above remarks apply more especially to the RED OR AMAZON ANTS, whose instinct for slave-making is marvellously

developed. It is generally at the decline of a warm summer
day that these ants, having made arrangements for a raid
upon some neighbouring colony, will array themselves in
serried columns, and start on their journey; their object
being to carry off the young ants, not yet fully developed,
and to make slaves of them; in fact, to make them perform
the duties which these plunderers are too lazy to attend to
themselves.

When these invaders reach the entrance of the nest they
intend to spoliate, they meet with considerable opposition by
the workers. A sharp contest takes place between them—
the one side bent upon plunder, the other determined not to
forego or lose either property or liberty, unless at the cost of
a well-sustained struggle. The amazons, being the stronger
party, succeed in conquering their opposers, and at once enter
the nest, commence their explorations, then seize the larvæ
and nymphs, which they carry off to mould to their own will.
The old ones of the nest they leave behind them, knowing,
no doubt, that it would be difficult to make them bend to
the yoke.

Every member of the invading party may be seen bearing
away the larvæ to their own little kingdom. No sooner are the
grubs safely lodged in the nest than the slaves already there
pay every attention to them. They feed, cleanse, and warm
the bodies of the young ants.

In the course of time, however, 'a change comes o'er the
spirit of the dream' of these pigmy warriors. Elated with
victory, flushed with success, and rendered careless and lazy
through the services or work done for them by the slaves,
they become indifferent to the changes which are gradually
taking place in the colony; and as they can neither build
their homes nor nourish their young, the whole scene changes,
their condition becomes a perilous one, they begin to feel
their own weakness, and that they are now dependent, for
future comfort, upon the very ants they had captured and
made slaves of.

These slaves, finding the nest too small, hit upon an emi-
gration scheme, which they carry into effect. They build a
larger nest, to which the amazons have to submit to be carried
by the mandibles of those who were once their captives, but
who are now their masters.

THE YELLOW ANTS are a match for the amazons; as they bid them defiance, frighten them by their mien, and their courage supplies the want of strength. It is said that one colony of ants will meet and fight with another near or midway between their respective domains. The battle-field is but small—seldom more than a square yard, but it presents many of the features of a human war. Great excitement prevails, and a determined effort is made on both sides to become the victors. Two and two will attack each other by their jaws. They will lock themselves together by means of their mandibles, and roll one over the other in the dust. The scene of the encounter is soon covered with the bodies of the dead and dying. Some of the ants may be seen carrying off prisoners to their nest; and although at night the contending armies beat a retreat, they often 'renew hostilities on the following day.' But for what do they fight? It may be for aphides, their *milch-kine*, or for the body of a beetle or fly, which are as valuable to them as our flocks and herds are to us.

Some ants having stings which they can use with considerable effect, and others having the power to send out from their small bodies a most irritating fluid which will raise blisters on the flesh it touches, they are not very desirable companions at a picnic party. Nevertheless it is known that

ANTS HAVE THEIR USES.—Rats, mice, and cockroaches have been known to quit a house when ants have entered it. Ants also teach useful lessons. Their affection for their young teaches us to value and to promote the happiness of posterity. Their obedience to their queen may teach us loyalty; from their economy we may learn prudence, and from their sagacity wisdom.

Dr. Gould says: 'A settlement of ants, the structure of the common workers, the character of the queen, the changes of the young, and the use they answer in the scale of beings, may teach us to admire the majesty of God, who has arranged the universe with so much beauty, and embellished each part of it with such a scene of wonders.'

CHAPTER VI.

FISH IN ARMOUR.

'Within the ever-rolling sea,
  Gliding its green waves through,
Are finny tribes from sprats to whales ;
Smooth skinned, or clothed with shining scales ;
And fish in armour too.'

WHEN the world was young, when mechanical genius was in its infancy, and working appliances were but few and of the rudest kind, it is natural to suppose that articles manufactured by man at that time would be coarse and defective compared with those of modern times.

A peep into a watchmaker's shop, especially at the complex mechanism of those very small watches, the turning round of whose tiny hands corresponds to the course of the sun and the rolling round of the planet in which we live, would almost lead one to suppose that man's mechanical genius and his taste for the delicate, the beautiful, and useful have nearly reached perfection. True as this might be, the organism of the many forms of life in the sea is far more wonderful than anything ever made by man.

If what has been stated be true, that naturalists have discovered more than 13,000 different kinds of fish, what a world of varied life, beauty, and wonders the sea contains !

A GLIMPSE AT THE FINNY TRIBES.—We have no intention to refer to the finny and vertebrated inhabitants of the deep, except in a general way, as it may relate to the marvellous adaptation of their organism to the element in which they live. After we have done this we shall describe, as briefly as possible, some of the shell-fish commonly eaten in our own country. How few people, comparatively, know anything of the peculiar manner in which our shell-fish are brought into existence, or of their singular structure, the changes they undergo, the particular localities of the sea in which they are found, or the way in which they are procured! Hence our reason for attempting the description which is to follow our remarks on the finny tribes.

It may be observed, that although the ocean contains the very largest as well as the very smallest of animals, they are all so organized as not only to live in the watery element, but no doubt to enjoy their existence in it.

In referring to the various and diversified inhabitants of the great deep, Rymer Jones observes, ' Many fishes are adorned by the hand of Nature with every kind of embellishment, variety in their forms, elegance in their proportions, diversity and vivacity in their colours; nothing is wanted to attract the attention of mankind.'

FISHES VIVIPAROUS.—Although the large majority of fishes are oviparous, there are some few whoproduce their young alive and perfect in form. The whale is one of them. She has one at a time, and, like animals of the mammalia family, suckles it from her own teats. The males of other finny tribes have the *milt*, and the females the *roe*. It is said, however, that some of the cod and sturgeon tribes have both. The spawn of some kinds of fish is deposited in sand or gravel, and some attach their ova to seaweeds. The fecundity of fish is supposed to be far greater than that of any other known race of animals.

Pike.

TEETH OF FISHES. —The teeth of fishes are very numerous, generally sharp-pointed, and firmly fixed in their jaws. In this respect the carp and pike are said to form an exception, as their teeth are so movable as to appear to be attached to the skin only. The pike, however, has teeth scarcely less formidable in size and sharpness than those of a good-sized dog. The tongue of fishes is in general motionless and obtuse. Being furnished with nostrils and olfactory nerves, as stated by some writers, fishes possess, in all probability, the sense of smelling. Although some doubt that these natives of the sea have the sense of hearing, it is asserted by Bingley ' that the organ of hearing is placed at the sides of the skull. In its structure this organ in fishes is by no means so complicated as in quadrupeds and other animals that live in the air.'

EYES OF FISHES.—As fishes have no eyelids, their eyes are in general furnished with a transparent skin that covers the rest of the head. This, with the cornea being flat, preserves the eyes from injury, and prevents them from wearing away in their quick passage through water. The eyes appear also to be supplied with a muscle, which serves to lengthen or flatten them as the animal may require. We believe it has never been positively ascertained whether fishes have the power to move their eyes in their sockets or not. Some fish, as the sole, have their eyes placed on one side of their head only.

FINS OF FISHES.—Different tribes of fishes may be distinguished from each other by their fins, which differ in number and in size. Some fins are *vertical*, and act as a keel or rudder. Those on the back are known as *dorsal* fins; those under the tail as *anals*, and at the end of the tail as *caudal fins*, which, being turned sideways, steer the fish as a rudder does a ship. The air-bladder, with which most of them are supplied, enables them to rise or sink in the water as they may think proper, and gives them that easy kind of buoyancy in the element they live in, which is not enjoyed by those who are without air-bladders, and have consequently to remain much at the bottom of the water.

FOOD OF FISHES.—Those kinds of fish found in fresh water live upon worms, the larvæ of water insects, molluscs, or flies that may alight on the surface. There can be no doubt that in many instances they are truly the scavengers of the water, and help as much in keeping it pure and sweet, as birds of prey and other carnivorous animals do in keeping the air free from the pernicious effluvia that would otherwise arise from decaying animal matter. The marine kind prey upon each other—the larger devouring the less, these living upon others less than themselves, and so on down to the smallest. It is a remarkable, interesting, and well-known fact that the smaller kinds of fish are not only more numerous and prolific than the larger ones, but that they can find food in shallow water near the shore, whither they cannot be followed by their pursuers, and so escape being devoured by them.

MIGRATION OF FISH.—Just as regularly as spring, summer, autumn, and winter come round, so 'shoals of herrings, sprats, cod, and haddocks visit or come near our shores at a particular time of the year, and quit them with equal regularity, without leaving a single one behind.' Just as regularly some kinds of fish visit certain places in which to deposit their spawn, and then, after so doing, return to the waters from whence they had come.

INTELLIGENCE OF FISH.—We have been informed by a gentleman that some fish kept in a large pond on his father's estate in Sussex have been taught to gather to any part of the pond at which they hear the ringing of a bell for the purpose of feeding. In this fact we have at least an indication of memory. These fishes were in the habit of associating the sound of the bell with the idea of satisfying their hunger. If there is memory there is reflection, and who can separate them from thinking? If fishes remember that a certain sound means to them a particular thing, must they not think?

While some people deny to fish the sense of hearing, others say, as already stated, that they at least possess the organ of it. It has been asserted that fish, possessing in a fair degree the sense of feeling, are guided or affected by those vibrations that are made by different sounds on the surface of water. Admitting this to be the case, and that they cannot hear, it still shows that fish not only possess intelligence, but enough of it to discriminate between vibrations produced by various causes. It may be reasonably supposed that had a pistol been fired about the same time over the pond referred to, the fish would not have collected as they did when the bell was rung, because they would not associate either the sound of the pistol or the vibrations caused thereby with a good feed of bread-crumbs or other food which they invariably did with the ringing of the bell. If fish then remember and think, they must have intelligence.

LONGEVITY OF FISH.—It is supposed that fish live to a very great age if not captured for food, or for any other purpose. Instead of suffering, as human beings and land animals do, from the effects of old age, their bodies seem almost continually to increase with fresh supplies of food; and as the animal grows, the conduits of life furnish their stores in greater abundance. It has been ascertained that some kinds of fish have arrived at 100, and even 150 years of age. This knowledge, however, has reference to those only that have been kept in private ponds, in which the numbers of these creatures are but small.

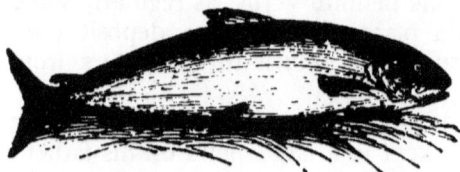

Salmon.

Although we sometimes hear of disease amongst salmon, and of other kinds of fish being affected in various ways, it is supposed that they are comparatively free from disease; indeed, much more so than those animals are that live on land.

The Rev. Mr. White, of Selborne, however, gives some information as to the manner in which fishes die, showing that in their native element they are sometimes attacked by unknown causes which deprives them of life. White says: ' As soon as a fish sickens, the head sinks lower and lower, and the animal stands as it were upon it, till becoming weaker, and losing all poise, the tail turns over, and at last it swims on the surface of the water with its belly upwards, by which time the life has ebbed out of it.'

FISH IN ARMOUR.—The reader will no doubt conclude that by ' fish in armour ' we refer to those inhabitants of the deep whose powers of locomotion are very feeble, but who are furnished with shells, more or less thick and strong, as a protection against the rapacious appetites of those larger and finny tribes who would otherwise devour their rich and pulpy bodies, and in time exterminate them from the world of waters.

These slow and almost motionless creatures belong to a class of *articulated* and *invertebrated* animals called *Crustacea* and *Mollusca*. Although we propose making some remarks on them, we do not intend to refer to all the numerous varieties which are included in the above two classes or divisions of shell-fish, but to those only which present great curiosities of structure, and especially those which are popular articles of consumption in our own country. We shall first consider those different kinds of shell-fish known as

CRUSTACEANS.—These creatures, which constitute, according to popular writers on natural history, the fifth class of articulated animals, present many wonderful and interesting points of structure, life, and habits. Jones remarks that 'Crustaceans are all oviparous. The female, after having laid her eggs, generally carries them about attached to the under part of her body, or sometimes enclosed in a sort of pouch formed of appendages variously modified ; sometimes the young undergo a very remarkable metamorphosis, and not only completely change their form during the earlier periods of their existence, but in the progress of their growth acquire additional limbs.'

In selecting and obtaining their food it is supposed that crustaceans are guided by some sense which answers to them the same purposes as does that of smell to the higher order of animals. There is some difficulty, however, in determining where this peculiar sense of shell-fish exists, whether it is connected with the first pair of antennæ, or with any other part of the body.

Although the sense of feeling is of the feeblest kind, yet observation corroborates the theory that shell-fish are not totally devoid of it. 'We have seen,' says a certain writer, 'a swimming crab hold its prey in one claw, while with the other it picked off morsel by morsel of the flesh, and conveyed it to its mouth in a manner which sufficiently evidenced the sensation of touch in these organs.'

THROWING OFF THE OLD COAT.—It is well known that periodically and at certain seasons changes take place in the coats and skins of most animal bodies. Horses and many other quadrupeds have yearly new coats of hair, while the fur of some of the smaller kind is of one colour in the summer and of another in winter. Toads and snakes change their skins ; birds moult, and new feathers replace the old ones ; and these fish in armour periodically cast off their old coats, or shells, and present themselves in armature perfectly new. There is something very wonderful in this moulting or change of shell in Crustaceans. The old one is not got rid of piece by piece, but altogether and unbroken, so as to present a counterpart of the perfect animal. We are informed that

6

every part of the integument is renewed; nothing is wanting in the cast-off skin, the antennæ, the jaws, and eyes are all there, every hair is represented by the case which enclosed it. Even the shelly plates from which the muscles originate, the tendons by which they are attached to the shell, the internal skin of the stomach, and the teeth which are hidden there, are found connected with the rejected shell !

The pressure of the old shell being removed, the animal suddenly increases in bulk, the new skin, as yet soft and flexible, allowing at first of great expansion; but it rapidly hardens, a stock of shelly matter having been for some time accumulating in its stomach in the form of two hard balls commonly called ' crab's-eyes.' This substance is supposed to be taken up and distributed to the surface, so that when the new crust has again acquired consistence these concretions are no longer found. The whole process occupies from one to three days. The supposition that the moulting in these animals takes place every year must probably be restricted to the period of their growth, beyond which the change of shell would seem to be unnecessary.

' Crustaceans,' says Figuier, ' vary greatly in colour; some are of a dark iron-grey, with a dash of steel-blue, like metal weapons forged for combat ; a few of them are red or reddish-brown, others are of an earthy yellow or of a livid blue. . . . They live on the sea-coast among the rocks and near the shore. Some few of them frequent the deep waters, others hide themselves in the sand or under stones, while the common crab loves the shore almost as much as the salt water, and establishes itself accordingly under some moist cliff overhanging the sea where it can enjoy both.'

Crustaceans are by no means epicures ; being carnivorous animals, nothing comes amiss to their voracious appetites. They can make a meal of any animal, living or dead, fresh or in a state of decomposition, providing it be large enough to satisfy their hunger. It may be their voracity has much to do with their merciless encounters with other animals, as well as with those of their own kind.

Some of these creatures are armed with powerful claws which, when engaged in warfare, they use unsparingly, and often with marvellous effect. They fight with each other sometimes for prey, or for a female ; and it often happens that both combatants are much injured thereby, especially in their antennæ, feet, and tails. Although in the course of time the mutilated members are replaced by new ones, the latter are usually small and weak. It appears evident from the above facts that if these subjects in armour show but little platonic love in their general habits, they can be irritated by jealousy, and also display in their fights with each other a good deal of courage, and we may say of martial ardour too. Indeed, do they not deserve the title of 'Naval warriors ?'

Having made some remarks on crustaceans generally, as

Crab.

constituting a distinct class or family of inhabitants of our seas, we shall now briefly refer to a few different kinds of them, especially to those that are articles of food in common use, and extensively identified with the commerce of this and of other nations. We shall notice first

THE CRAB FAMILY.—Crabs appear to belong to the section of ten-legged, short-tailed crustaceans. They are covered by a very strong calcareous shell. Although to superficial observers the organs of vision are not perceptible, it is said that the sense of sight in most of the species is peculiarly acute, and enables them to distinguish objects from a considerable distance. But they are most remarkable for a complex and elaborate apparatus for mastication. The mouth is furnished at least with eight pieces or pairs of jaws, which pass the food through an extremely short gullet into a membranous stomach of considerable size.

There are many kinds of crabs, distinguished from each other by peculiarities of structure and habits. The following are included:

Swimming Crabs.—These live at a distance from the shore, and often make for the high sea, being aided in so doing by having each foot furnished with a flat joint resembling a fin, with which they propel themselves rapidly through the water.

Spider Crabs.—These creatures, as their name implies, have long legs, but they are slow in motion. They are usually found amongst rocks and stones, covered with seaweeds, over which they can stride with comparative ease. At times, these crabs are covered with seaweeds, sponges, and other marine productions. During the winter they become torpid, and lie concealed in the crevices of rocks, or buried in the soil.

Shore-Crabs are well adapted in their structure for walking; that duty they perform by using their eight hind-legs only, which they do with so much facility as to be able to walk forwards, backwards, or sideways, just as they please. They are also good climbers.

Racer Crabs.—These are found on the coasts of Syria, Barbary, and the West Indies. They derive their name from the rapid pace at which they can travel, especially when threatened by danger.

Beckoning or Calling Crabs are so named because of a habit they have of flourishing one of their large claws about as if beckoning to some one at a distance. They make burrows in the seashore, where they remain a great deal of their time, hardly ever leaving them except to visit the sea, in which they deposit their eggs.

Hermit Crabs are so called because, being very defenceless, they seek protection by getting under an empty shell of some other kind of fish, which they carry about.

Edible Crabs are those commonly known at our tables. They are nocturnal in their habits. They often visit the shore in search of food, and seize on anything suitable to their palate. Should they remain too long in their hunts after food the waters may retreat, and thus leave them stranded, exposed to danger, and subjected to considerable inconvenience. They will, however, look out for a hole, or some corner into which they will squat, and patiently wait for the return of the tide.

'Crab-fishing,' says Jones, 'is usually conducted by two men, in a boat provided with lines and creels, cruives, or crab-pots, made of a kind of osier basket-work. They are constructed upon the same principle as a wire mouse-trap, but the aperture, instead of being on the side, is at the top. The bait, which consists of stale fish, is fastened to the bottom, and the creel is then sunk in a favourable situation by stones of sufficient weight within it; a line is fastened to the creel, to the upper end of which a cork is attached. The bait can readily be seen by the crabs, which on entering are caught like rats in a trap; the difficulty of egress being increased by the aperture being overhead.'

We may here notice that the floats indicate to the crab-fishermen the precise position of the crab-pots, which at certain times the men visit, and which, if a sufficient number of crabs is ascertained to be within, they haul up to their boats, put them into well-boxes, and there keep them until they are sold.

CRAY-FISHES.—These inhabitants of the sea bear a resemblance to the common lobster, both in shape and flavour. Their antennæ are large and covered with spines. Their shell is rough, and abounds with prickles. They prey upon the smaller kinds of fish, and any marine production that may be suitable for them. They live in very rocky places, where fishing is rather difficult. In consequence of this, cray-fishes pass an almost undisturbed life; and it is said many of them live to a great age, and others grow immensely large, even three feet long. They are caught in wicker-baskets baited with flesh.

LOBSTERS. — Lobsters, like crabs, constitute a popular article of diet not only in our own country but in many others. They are abundant in the Mediterranean, throughout the European Seas, and on the eastern coast of North America. When alive they are of a dull, blue colour. It is by boiling them that their shells become red. Many of the lobsters consumed in England are brought from various parts of our own coasts, and some from Norway. The females produce an almost incalculable number of eggs. 'Upwards of twelve thousand have been found under the tail of a single individual. Lobsters periodically cast off their old shells when they have become too small for the increasing size of the internal body. It is supposed by some writers on shell-fish that to accomplish this object these animals observe in some quiet place a season of fasting, so that the inside of the limbs may shrink a little to permit being drawn through the narrow joints. Be this as it may, it is well-known that they do get rid of the old shell, and that a new one appears in its place, which, being of larger dimensions, allows room for the growing body. By a stroke of their tail lobsters can spring backwards twenty or thirty feet.

Lobster.

SAGACITY OF A LOBSTER.—We are informed by *Nature* that 'A few days ago, at the Rothesay Aquarium, a tank containing flat-fishes was emptied, and a flounder of eight inches in length was inadvertently left buried in the shingle, where it died. On refilling the tank it was tenanted by three lobsters (*Homarus marinus*), one of which is an aged veteran of unusual size, bearing an honourable array of barnacles; and he soon brought to light the hidden flounder, with which he retired to a corner. In a short time it was noticed that the flounder was *non est*. It was impossible the lobster could have eaten it all in the interim, and the handle of a net revealed the fact that, upon the approach of the two smaller lobsters the larger one had buried the flounder beneath a heap of shingle, on which he now mounted guard. Five times within two hours was the fish unearthed, and as often did the lobster shovel the gravel over it with his huge claws, each time ascending the pile and turning his bold defensive front to his companions.'

This story seems to prove that lobsters possess, in some degree, the sense of smell, or how could the one referred to have known the dead fish was there, as it was hidden from sight? It is also evident the lobster showed great tact and intelligence, for after feeding on the flounder, he not only re-buried it, but, to make sure of not being robbed of it, acted as sentinel over the larder he had made under the shingle.

PRAWNS.—These marine crustacea, during the summer-time, exist in great numbers at the mouths of the rivers of our own country, and are caught in nets formed like sacks fastened on the ends of long poles. These are thrown out to sea as far as possible, and then drawn to the shore, when they are generally found to contain great numbers of these small crustaceans. Like the other members of the same family, prawns cast off their old coat or covering, which is to give place to another more suitable for the requirements of the animal, both as to room and to comfort. During this change of coat the prawn never feeds, but is uneasy, and moves from place to place until it meets with a locality suitable for its purpose.

When this has been found, the prawn stretches out its legs, then hooks its feet to some substance sufficiently firm to hold by. It then sways its body backwards and forwards, by which means it loosens the entire surface of the body from the carapace; by a little more swaying and gentle effort the animal gets rid of its old covering. The prawn, thus liberated, rolls itself on the ground, and there lies perfectly soft and helpless. In time, however, a change takes place, the creature gains strength, and assumes a fresh appearance, looks beautiful in its new dress, and becomes as lively, bold, and active as ever.

SHRIMPS are allied to the lobster, and are not more than two inches long. They are found in shallow waters along our coasts, and are secured by dredge-nets. They have ten feet each; move about by leaps, and constitute food for other fishes.

MOLLUSCS.—These fish in armour comprise a very large number of different kinds. They are arranged into seven separate classes, some of which are divided into orders and families. We shall, however, refer more particularly to those included in the third class.

Speaking in general terms of this large division of invertebrate animals, we may state that they are without bones, have no internal skeleton, and may be defined as soft, fleshy animals. Some kinds live on land and breathe the air. They are found in our gardens, fields, on the branches and trunks of trees, on the banks of our hedges and in our plantations. Some live in fresh water, others at the bottom of streams, and some are amphibious. Vast numbers are found in the seas of almost every part of the world.

According to 'Animal Creation,' the number of species already in museums reaches from 8,000 to 10,000. There are cabinets of marine shells, bivalve and univalve, which contain from 5,000 to 6,000, and collections of land and fluviatic shells which count as many as 2,000. The total number of molluscs, therefore, probably exceeds 15,000 species.

From what we have stated, it will be seen that these creatures constitute very wonderful links in the chain of being. This will appear to be more so when we remember that many of them are food for other animals than man, as well as supplying valuable materials for the fine arts.

CLAMS.—This family of shell-fish demands a short notice, principally on account of their large size, and the novel uses to which their shells are applied. Rymer Jones says they are the giants of the bivalve race. They live, attached by their byssus, to rocks, shells, and corals. The valve of a large individual forms a very picturesque basin for catching the clear falling water of a fountain, which flows prettily through its deeply indented edges. In Roman Catholic countries the valves of this huge shell are sometimes employed as *bénitiers*, or vessels for containing holy water. . . The byssus is so thick, and its attachment to the rock so strong, that it is frequently necessary to cut it with a hatchet in order to obtain the animal. Clam-shells are sometimes more than two feet across, being the largest known.

COCKLES, though small in size, are an important article of food, and exist in great numbers on sandy shores. Their organization is wonderful. Improbable as this may appear to those not well acquainted with this kind of shell-fish, it never-theless is provided with one foot of good size, compared with the rest of its body, which it uses for burrowing. This foot assumes different shapes, according to the use the cockle wishes to make of it. When the animal desires to bury itself, the foot is lengthened into a wedge, which it thrusts down into the sand, and then, turning the end into the form of a hook, pulls itself down into the hole made by its wedge-foot. In this condition it will remain for some time, breathing all the while through some projecting tubes which are just visible. By bending the end of the foot, and pushing against the sand with great force, the shell is again extruded. When in the water, the cockle often moves rapidly along by pushing its foot against the ground, just as a man would propel his boat along by using his oars on the surface of the water.

MUSSELS.—According to the 'Students' Natural History,' mussels constitute a family of bivalve mollusca. They are marine animals, and are found attached to rocks and stones by a beard of stout fibres. They are common in this country, and on the coasts of the Mediterranean and the North Sea. They frequent mudbanks which are uncovered at low water, and immense quantities are used for bait by fishermen. About fifty species have been described distributed over all the world.

The shell of the eatable mussel is of oblong shape, and of a dark purple colour. Like the cockle, the mussel is furnished with a kind of foot, capable of retraction and elongation, which is enveloped in a sheath of fibres. By employing these fibres the animal can weave a number of silky threads by which it can fasten itself to the rock. Some people deem mussels to be poisonous, as painful effects have been produced by eating them. The precise cause, however, has never, we believe, been satisfactorily explained. Some writers suppose that as these animals are very indigestible, all persons who have weak stomachs must of necessity suffer by consuming them, especially if too freely. It has also been stated that the poisonous effect is produced by certain small crabs, which lodge themselves as parasites in the shell of the mussel.

The Pearl Mussel is worthy of special notice, as being the producer of the 'mother-of-pearl' and pearls. The cutting and polishing of the shells of these animals gives employment to great numbers of people, 'mother-of-pearl' produced thereby being an important article of commerce; and very valuable chairs, tables, workboxes, trays, and other things are inlaid with it. Pearls have also been found in these shells, differing, however, in colour, form, size, and lustre, and also in value.

The most extensive pearl-fisheries are those of the Persian Gulf and of the island of Ceylon.

OYSTERS.—The oyster family embraces several genera. Beard says, 'The best known, and perhaps the most important species of the genus is the common oyster, *ostrea edulis,* forming a considerable article of commerce. This mollusc is a very abundant one on the British coasts, and is generally found in banks or beds, several fathoms under water, and often extending to a considerable distance. The time of spawning is in May and June, and the fry, which are called 'spats,' are collected in vast numbers, and removed to artificial grounds or tanks where the water is shallow. These oysters are called 'natives,' and require five or seven years to attain their full size, while the 'sea-oysters' become fully grown in four years.

Oyster-fisheries are found along the coast of Essex and Kent, and on the coast of the island of Jersey, at which great numbers of men, women, and children are employed. From these and other places tens of thousands of *native* and *sea-oysters* are sent every year to the markets and shops of London, and to those of other towns.

Oysters are secured by a dredge net, edged with a scraper of iron which is drawn along the bottom by a rope attached to the boat. These bivalves have many enemies which constantly try to obtain them for food. They have, therefore, to be watched and guarded with jealous care, especially against star-fishes, which would otherwise be very destructive to them during their breeding-season.

The value, or rather the price, of oysters, like that of other kinds of animal food, has increased amazingly during the last few years. They are at present about four times dearer than they were forty years ago. The reason is no doubt owing to the extension of oyster culture not keeping pace with the increased demand of a rapidly growing human population.

PEARL-OYSTERS are contained in shells which are included in what are known as 'pearl-shells.' Pearl oysters, like mussels, produce mother-of-pearl, which is a beautiful substance, and much used in inlaying cabinet-work, making paper-knives, and many ornamental articles.

In referring to these oysters Rhind says : 'Pearls are roundish bodies found attached either to the inside of the shells of this species, or loose between the shells. Pearls the size of a pea are worth about a guinea, and when of the size of a peppercorn, about two shillings each. Small pearls are called 'seed-pearls,' and are of much less value. The famous pear-shaped pearl belonging to the Shah of Persia cost £10,000.

Pearl-oysters are obtained by men who dive for them, often in about twelve fathoms of water. This work is very arduous, unwholesome, and dangerous. It is well known that every year very many men who follow this employment lose their lives, in consequence of which many a widow is deprived of her son, and many a wife and family of their only support. Truly we may say, What mystery lies within our differing fates !

SCALLOPS. —In some respects scallops resemble oysters, especially in their external appearance, but they differ some-what internally. They have a moderate-sized, hatchet-shaped foot, which helps them to move about both by land and water. When they have been left on the shore by the receding tide their singular style of locomotion in attempting to regain the water is both amusing and interesting. They will open their shells widely, and then, shutting them with a sudden jerk, throw themselves forward, and so on until they reach the sea. These creatures often assemble in troops when the sea is calm, on the surface of which they glide gracefully along, having the upper valve of their shells open, which look like numbers of tiny sails moving on the water. When they apprehend danger, they close their shells and suddenly sink to the bottom.

It is said these bivalves have eyes which appear on the margin of their mantle as so many bright spots. Although there is some doubt about this, it is well-known that many of the species have finely coloured handsome shells. The shell is of considerable size, and Baird says : 'The deeper valve was used formerly as a drinking-cup, and celebrated as such in Ossian's 'Hall of Shells.'

'The St. James's shell, common in the Mediterranean and nearly allied to the scallop, was worn by pilgrims to the Holy Land, and became the badge of several orders of knight-hood.'

CHAPTER VII.

FIRST COUSINS, OR OUR BIRDS IN BLACK.

First cousins are we all as black as a coal,
So we can't disagree on that matter at all;
Although many think they've a right to abuse us ;
And wrongly assert that we haven't our uses :
This grave charge against us, allow us to say,
Is false, as we labour like slaves all the day
To pick up our living ; ' deny it who can.'
We also do good both to Nature and Man.

IN the remote, dim ages of the past, animals constituted themes for writers of both sacred and
profane history. Modern poets and painters
have so represented and popularized many of
them in verse and on canvas, that those who
study the characteristics of animals portrayed in the manner
referred to may learn many interesting, important, and useful
lessons therefrom. The raven and jackdaw have been immortalized in song ; whilst the crow and the rook have formed
subjects of study both to natural historians, physiologists, and
psychologists in almost every age and country. Thus it is
that animals have been brought prominently into notice.

To what has already been written about animals we venture
to add a small amount of information on the structure, habits,
life, and uses of those animals we denominate ' First Cousins,
or Our Birds in Black.'

The colours of birds are often suggestive of certain customs
observed by mankind. When we see a swan, a white goose,
or duck, we think of the Chinese custom of wearing white
dresses at their funerals. Birds in brown or grey colours
remind one of the dress of our friends the Quakers ; and

without wishing to be invidious or irreverent, our 'birds in black,' not in their habits, but in their colours, remind us not only of the sombre aspect of our funerals, but of those gentlemen who are our preachers and guides in matters of morality and religion.

Quadruped, bird, and insect life help very materially to render our landscapes charming, attractive, and really enjoyable. During the greater part of the year, Nature gives us almost endless forms of life in our fields, woods, and hedgerows, in which we have bird-life wild, free, and happy, the little winged workers of Nature ever chirping, chattering, or singing, as they perform their daily task for self support, or seek food for their young and helpless progeny. Two or three representatives of bird-life will suffice for this chapter.

THE ROOK.—This name is very suggestive of country mansions, clusters of fine old elm-trees, of extensive parks, and baronial estates. Rooks claim relationship with the *conirostral* tribe of the *Insissores*. They are gregarious, and live all the year round in societies. They sometimes keep company with jackdaws and starlings. As many as 20,000 rooks have been known to live in one rookery. It may often have been a matter of wonder that their nests

Rook.

or cities built on high, with their black inhabitants, have not, when the wind has blown violently, all tumbled down together. But rooks, like other birds, not only know *when* and *where* to build, but *how* to build to make their homes secure.

Although as a rule rooks constitute a happy family, now and then their peace is interrupted, and discordant elements are sometimes at work among them. They are very jealous of strangers who may try to intrude themselves in a rookery to which they do not belong. This appears to produce among the old proprietors of the nests in such a rookery great excitement and indignation; so much so, as some have stated, that they assemble a rook parliament for consultation as to the best mode of punishing the intruder.

Be this as it may, it is well known that if a rook is too lazy to build a nest for itself, and takes possession of another, the act is noticed by other members of the colony, especially by the owners of the nest, who give it timely notice to quit. If this is not done quietly and quickly, the invader is ejected by main force, sometimes at the cost of his life.

Occasionally young members of the rook family attempt to build nests outside the limits habitually established, on which the other members of the republic destroy their handiwork as soon as formed, and enforce conformity to ancient custom and the will of the majority. White says, 'Some unhappy pairs are not permitted to finish any nest till the rest have completed their building, and that sometimes they pull each other's nests to pieces.' The same author informs us that 'as soon as rooks have finished their nests, and before they lay, the cocks begin to feed the hens, who receive that bounty with a fondling tremulous voice and fluttering wings, and all the little blandishments that are expressed by the young while in a helpless state. This gallant deportment of the males is continued through the whole season of incubation.'

Markwick states that 'After the first brood of rooks is sufficiently fledged they all leave their nests in the daytime, and resort to some distant place in search of food, but return regularly every evening, in vast flights, to their nest-trees, where, after flying round several times with much noise and clamour till they are all assembled together, they take up their abode for the night.'

Rooks are furnished with bills well adapted in length, strength, and shape for digging in the soil, in which they find their food. Although they may consume newly sown seed or ripening grain, they are useful to the farmer in checking the too rapid increase of the grub of the cockchafer, red and white wireworms, and other forms of destructive insect-life.

The rook has been confounded with the crow on account of the similarity in their appearance. Rooks live in flocks— crows in pairs, and the latter also feed upon carrion and small animals, which rooks are not known to interfere with. Rooks exhibit considerable forethought in repairing their nests during the autumn, as if they knew this was necessary to secure them against the high winds and storms of the coming winter. These birds are, by Act of Parliament, protected during the close, or breeding season.

NOTICE TO QUIT.—Rooks are sagacious. Some very old elm-trees, in which was a rookery, not far from London, were condemned to be cut down and young ones planted in their place. The oldest of the trees were first doomed, and a piece of their bark taken off to indicate their fate. These trees were soon forsaken by the rooks, and it was afterwards observed that when the other elms were marked in a similar manner the rooks at once forsook the trees, as if fully aware that the removal of the bark was a notice to quit.

THE GUN AND THE CRUTCH.—The rook has been known to fly from a man carrying a crutch on his shoulder, and yet to endure the approach of the same man when he walked with a limping gait with the crutch under his arm.

THE LAWYER'S ROOKERY.—About the commencement of the Volunteer movement there existed a rookery in a part of London much frequented by gentlemen learned in the law. All went on very pleasantly with the rooks until permission was granted to one of the Volunteer corps to drill in an open space of ground contiguous to the rookery. The birds did not object to the tramping, the manœuvring, and other et-ceteras of military training, but they were not disposed to tolerate the sound of the bugle which called the men to drill, so they forsook the locality entirely, and remained away for some time. Their departure was so much regretted by the property-holders and renters of chambers in the place referred to, that at their request the drill was discontinued. At the following season the rooks returned to their old quarters, again occupied their nests, in which they reared their young, and there they have remained ever since, no doubt well pleased and grateful to those gentlemen who appreciated their *caws*, and did not object to their company.

The poet Crabbe thus speaks of these birds :

'Rooks unnumbered build their nests—
Deliberate birds, and prudent all ;
Their notes, indeed, are harsh and rude,
But they're a social multitude.'

SYMPATHETIC FEELING OF ROOKS.—The following singular example of affecting sagacity and social feeling by which rooks

are characterized is mentioned by Dr. Percival in his 'Disser-
tations': 'A large colony of rooks had subsisted many years
in a grove on the banks of the river Irwell, near Manchester.
One serene evening I placed myself within view of it, and
marked with attention the various labours, pastimes, and evo-
lutions of this crowded society. The idle members amused
themselves with chasing each other through endless mazes,
and in their flight they made the round with an infinitude of
discordant noises. In the midst of these playful exertions it
unfortunately happened that one rook, by a sudden turn,
struck his beak against the wing of another. The sufferer instantly fell into the river. A general cry of distress ensued.
The birds hovered with every expression of anxiety over their
distressed companion. Animated by their sympathy, and
perhaps by the language of animals, known to themselves, he
sprang into the air, and by one strong effort reached the point
of a rock which projected into the river; the joy became loud
and universal; but alas! it was soon changed into notes of
lamentation, for the poor wounded bird, in attempting to fly
towards his nest, dropped again into the river, and was
drowned, amid the moans of his whole fraternity.'

THE RAVEN.—Although very many birds are mentioned
in the Scriptures, there are but few of them more prominently
referred to, or more intimately associated with many of the
important events recorded therein, than the *raven;* a fact
which invests the study of this bird with more than ordinary
interest. At the time of the Deluge Noah sent out of the
ark a *raven*, which went to and fro, until the waters were dried
up from the earth. When God commanded Elijah to hide
himself by the brook Cherith, He said: 'And it shall be that
thou shalt drink of the brook; and I have commanded the
ravens to feed thee there. . . . And the *ravens* brought him
bread and flesh in the morning, and bread and flesh in the
evening; and he drank of the brook.' *Ravens* are specially
spoken of as being the objects of God's care: 'He giveth to
the beast his food, and to the young *ravens* which cry.' These
birds are also represented as agents employed to punish the
wicked: 'The eye that mocketh at his father, and despiseth
to obey his mother, the *ravens* of the valley shall pick it out,
and the young eagles shall eat it.'

Ravens belong to the same family as the rook, and are

found in almost every part of the world. Their general colour is black, finely glossed with blue. 'They are bold, keen, and sagacious.' As they live partly on carrion, fruit and insects, their beaks are intermediate between that of the vulture and woodpecker. Their bills are straight, so that they can inflict a severe wound by thrusting; they are a little hooked at the end, by which they can keep a firm hold on anything and tear it to pieces. They defend their nests against eagles and vultures in a very vigorous manner, but not against man.

Many ravens are two feet long; they usually mate for life, and some of them live to be a hundred years old. They live in pairs, and build their nests in high trees, sometimes in rocks. They brave the rigour of an Arctic winter; but some of them travel southward when the weather is severe. Like jackdaws, magpies, and parrots, ravens may be taught to articulate words and short sentences, which they sometimes utter with great distinctness and emphasis. Lee says: 'They speak so plainly that they have more than once been known, by uttering exclamations of surprise and alarm, to turn out a guard of soldiers when placed in the neighbourhood.'

DOMESTIC RAVENS, when taken very young, become exceedingly amusing. Some of their characteristics are remarkable and interesting. These birds are, however, generally busy, inquisitive, and impudent; they go where their fancy leads them; they affront and drive off the dogs, play pranks on the poultry, and are particularly careful to ingratiate themselves into the favour of the cook, who with them is a much-loved and valued friend.

RAVENS AS EVIL OMENS.—In times of ignorance, these birds were looked upon with apprehension. This, no doubt, arose from their black colour, disagreeable croak, and fetid odour. Linnæus says, 'that in the southern provinces of Sweden ravens have been seen soaring along amid a thunderstorm and electric fires, which seemed to stream from their bills, enough to terrify the ignorant and superstitious, and to make them stamp these birds with the attributes of a demon.'

We are informed by Jesse that 'A sober hind at work in a certain neighbourhood stated that his companion had been warned of his approaching death in consequence of a raven having always croaked when it flew over his head.'

In Eastern countries the raven has been held in great veneration as being the bird who fed the prophet in the wilderness.
Even the Romans, who thought the bird ominous, paid it,
from motives of fear, the most profound veneration.

How RAVENS CLING TO THEIR EGGS.—If birds and other
animals are bold in defence of their offspring, which they
almost invariably are, they are equally tenacious in clinging
to their eggs during the time of incubation. We are informed
by one writer that he once lifted a hen blackbird off her
nest, but that she came back when he had removed a few
feet away. Referring to the devotion birds exhibit while
sitting on their eggs, White gives the following remarkable
instance. He says, 'that a small wood called Losels was
furnished with a set of oaks of peculiar growth and value. In
the centre stood a tree which, though stately and tall on the
whole, bulged out into a large excrescence about the middle
of the stem. On this oak a pair of ravens had fixed their
residence for such a series of years that the oak was distinguished by the title of the Raven Tree.

'Many were the attempts of the neighbouring youth to get
at their eyrie; the difficulty whetted their inclinations, and
each was ambitious of surmounting the arduous task. But
when they arrived at the swelling, it jutted out so in their
way, and was so far beyond their grasp, that the most daring
lads were awed, and acknowledged the undertaking to be too
hazardous; so the ravens built on, nest upon nest, in perfect
security, till the fatal day arrived in which the wood was to be
levelled.*

'It was in the month of February, when these birds usually
sit. The saw was applied to the butt, the wedges were
inserted into the opening, the woods echoed to the heavy
blow of the beetle, mall, or mallet, the tree nodded to its fall;
but still the dam sat on. At last, when it gave way, the bird
was flung from her nest, and though her parental affection
deserved a better fate, was whipped down by the twigs, which
brought her dead to the ground.'

A FAMILIAR RAVEN.—'In October, 1822, there was in the
possession of Mr. James Weymess, the gamekeeper at Riddle-

---

* It appears that wood was required to repair the bridge at the Toy,
near Hampton Court. Twenty of the finest trees were used for this
purpose.

ham Hope, the seat of Charles John Clavering, Esq., a young raven, fifteen months old, which was taken from the nest when very young, and brought up by the keeper with the dogs. It was so completely domesticated that it would go out with the keeper and the dogs, and when it took its flight farther than usual, at the sound of the whistle it would return and perch upon a tree or a wall and watch all their movements. It was no uncommon thing for it to go to the moors with him, and to return, a distance of ten or twelve miles. It would even enter a village with the keeper, partake of the same refreshment, and never leave him until he returned home—a circumstance, perhaps, never yet recorded in the annals of natural history.'

THE RAVEN RATCATCHER.—' A gentleman in Perthshire brought up and kept a tame raven in his stables, which proved of great use in destroying rats, and this he performed with a degree of cunning and adroitness which could scarcely be exceeded by human intelligence. The time he fixed on for his work of destruction was generally in the forenoon, when the servants were out airing the horses. On such occasions Jacob (this was the raven's name) took care to provide himself with a bone on which there was some meat, and this he placed opposite the rats' holes in front of the crib, and then perched himself above, watching with a steady and clear look the spot where the bone was laid. This bait seldom failed to attract the scent of the rats when all was quiet, and no sooner did they make their appearance than he darted down on them, and seldom missed his aim ; and having seized them, they were despatched in an instant. And what was singular, he did not eat them when at first secured, for he generally carried them to the sill of a window, returning to the sport, in which he seemed to take great interest. And he has been known to kidnap half a dozen in a forenoon. When his sport was interrupted by the return of the horses, he carried off his booty, one by one, to a neighbouring tree, where there was an old crow's nest, in which he deposited the spoil, and fed on them at his leisure.'

THE RAVEN AND GLOVES.—' Many years since a man in Sunderland, being employed in hedging near to an old stone quarry, went to eat his dinner in a deep excavation in order to be sheltered from the weather, which was very stormy, and as he went along, pulled off his hedging-gloves and threw them down at some distance from each other. While at his repast, he observed a raven pick up one of them, with which it flew away, and very soon afterwards returned and carried off the other. The man, being greatly surprised, rose to see if he could trace where the bird had gone with his gloves. He scarcely had cleared the quarry before he saw large fragments fall down into the very place where he had been seated, and where, if he had continued a minute longer, he must inevitably have been crushed to pieces.'

THREE FUNNY RAVENS.—'A raven called the "Parson" lived in a stable, and observed that when the groom tickled his favourite horse behind the shoulder, the hind-legs of the horse would go up, and the groom would frequently say to the playful animal, "Ha, Jack! go it, old fellow!" One day the groom heard capers and noise in the stable, and approaching the door, was astonished by the sound of his own voice inside—"Ha, Jack! go it, old fellow!" Upon entering the stable, he found the raven perched on the horse's hind quarters, pulling hairs out of Jack's tail, and responding regularly to the horse's kicks with the groom's exclamation, "Ha, Jack! go it, old fellow!"

'Another, addicted to peeping and listening from the porch of a chapel during divine service, heard the minister say repeatedly, "Let us pray." On the occasion of a tea-party at the Squire's Hall, this animal was brought into the drawing-room to amuse the assembled company with his tricks. Set down, he looked about him, nothing abashed. Presently kenning an old well-known antiquary, clad in a dark snuff-coloured suit, whose head and shoulders only were visible over the top of a high-backed chair, on which the worthy man happened at that very moment to be kneeling, our sable hero, assuming a solemn attitude, gave out slowly, and in pulpit tones, "Let us pray." Another, who set up his abode at a posting-house in Yorkshire, guarded the yard with the fidelity of a watch-dog, and upon the arrival of a traveller, invariably demanded with a loud voice, "Ostler, come and take the gentleman's horse!"'

MRS. GRIMALKIN TONGUE-TIED.—A writer in the *Workmen's Messenger* says : 'Some years ago I was walking along a retired street, when, hopping at some distance before me, I saw a raven, which had evidently strayed from its owner. I ventured, though with some misgivings, to stretch out my hand to him; immediately, with a hoarse croak, he jumped on my wrist, and turned his head sideways to get a good look at me. Having satisfied himself as to my respectability, he settled himself on my wrist, murmuring his pleasure in a series of jerking sounds.

'The next day I took him out in the garden to give him a walk. As he marched solemnly along before me, evidently meditating on the changes of life, a large tabby cat, a great pet of mine, sprang suddenly upon him. Being a town-bred cat, she was under the mistake that he would be as easily disposed of as a sparrow. What was her horror and dismay when, adroitly thrusting his beak into her open mouth, Jack seized her by the tongue! Poor pussy ran along the path, struggling to get free, while he hopped beside her, flapping his wings exultingly, until I came to the rescue. Not pitying Mrs. Grimalkin much for the fright she had got, I hoped it would be a warning to her not to interfere with birds in future, she having already, in her love for dainties, demolished several of my pet canaries.

'From this time Jack generally paid her a visit once a day, much to her dislike, especially as she had a family of young kittens. Jack would watch his opportunity, and when she least expected it, his roguish eye and shining head would suddenly appear before her startled gaze.

'With a spring he would quickly poise himself on the edge of the basket, and after apparently pausing to make a selection, with a sudden dive he would snatch up a kitten, hold it suspended for a moment, and then let it fall back into its place, enjoying with mischievous delight the agony of the mother while he had possession of her offspring.

'Jack was an inveterate thief, and hid quite a store of little articles under the edge of the carpet, as a fruit-knife, a pair of scissors, a thimble, etc.'

The *Animal World*, referring to CHARLES DICKENS'S RAVENS, says: 'It may not be out of place to copy, for the perusal of our youthful readers, the first part of the preface to "Barnaby Rudge," which gives us a glance at the interest taken by the great author in one member of the great family of "Our Feathered Companions." Who has not read of

"Grip"? "As it is Mr. Waterton's opinion that ravens are gradually becoming extinct in England, I offer a few words here about mine. The raven in this story is a compound of two great originals, of whom I have been, at different times, the proud possessor. The first was in the bloom of his youth when he was discovered in a modest retirement in London, by a friend of mine, and given to me. He had from the first, as Sir H. Evans says of Anne Page, 'good gifts,' which he improved by study and attention in a most exemplary manner. He slept in a stable—generally on horseback—and so terrified a Newfoundland dog by his preternatural sagacity, that he has been known, by the mere superiority of his genius, to walk off unmolested with the dog's dinner from before his face. He was rapidly rising in acquirements and virtues, when, in an evil hour, his stable was newly painted. He observed the workmen closely, saw that they were careful of the paint, and immediately burned to possess it. On their going to dinner, he ate up all they had left behind, consisting of a pound or two of white lead; and this youthful indiscretion terminated in death. While I was yet inconsolable for his loss, another friend of mine, in Yorkshire, discovered an older and more gifted raven at a village public-house, which he prevailed upon the landlord to part with for a consideration, and sent up to me. The first act of this 'Sage' was to administer to the effects of his predecessor, by disinterring all the cheese and halfpence he had buried in the garden—a work of immense labour and research—to which he devoted all the energies of his mind. When he had achieved this task, he applied himself to the acquisition of stable language, in which he soon became such an adept, that he would perch outside my window and drive imaginary horses with great skill all day. Perhaps even I never saw him at his best, for his former master sent his duty with him, 'and if I wished the bird to come out very strong, would I be so good as to show him a drunken man,' which I never did, having (unfortunately) none but sober people at hand. But I could hardly have respected him more, whatever the stimulating influences of this sight might have been. He had not the least respect, I am sorry to say, for me in return, or for anybody but the cook, to whom he was attached; but only, I fear, as a policeman might have been. Once I met him unexpectedly, about half a mile off, walking down the middle of the public street attended by a pretty large crowd, and spontaneously exhibiting the whole of his accomplishments. His gravity under those trying circumstances I never can forget, nor the extraordinary gallantry with which, refusing to be brought home, he defended himself behind a pump, until overpowered by numbers. It may have been that he was too bright a genius to live long, or it may have been that he took some pernicious substance into his bill, and thence into his maw, which is not improbable, seeing that he new pointed the greater part of the garden-wall by digging out the mortar, broke countless squares of glass by scraping away the putty all round the frames, and tore up and swallowed, in splinters, the greater part of a wooden staircase of six steps and a landing ; but after some three years he too was taken ill and died before the kitchen fire. He kept his eye to the last upon the meat as it roasted, and suddenly turned over on his back with a sepulchral cry of ' Cuckoo !' Since then I have been ravenless."

' The great author himself is now silent in death, but his affection towards all living creatures, " man and bird and beast," still lives and endears his memory to many thousands of mankind ; for in this respect, like his own Barnaby, " he was known to every bird and beast about the place, and had a name for every one of them." But we are reminded also, when we look upon the raven, of Southey, Byron, Longfellow, and Poe, who have told us something of the pranks, virtuous and vicious, of this clever bird ; and of Goldsmith also, who says he " heard a raven sing the ' Black Joke ' with great distinctness, truth, and humour." There can be no doubt about the linguistic powers of our hero throughout all ages, for he is at the head of talking birds.'

CROWS.—These birds seldom associate in flocks, but for the most part remain in pairs. The crow is very much like the raven in form and habits, but much smaller. The bill is more curved than the rook's, and its voice is hoarser. They feed upon whatever comes in their way. Montagu says that he saw one pursue a pigeon, on which it pounced like a hawk, and another that knocked a pigeon dead from a barn-floor.

There is a story told that seems almost incredible. A crow, perceiving a brood of young chickens, fourteen in number, under the care of a parent hen, picked up one of them; but a young lady, seeing what had happened, suddenly pulled up the window, and calling out loudly, the plunderer dropped his prey. In the course of the day, however, the audacious and calculating robber, accompanied by thirteen others, came to the place where the chickens were, and each seizing one, got clean off with the whole brood at once. Clearly they must have a language by which

Crow.

to carry out such a plot as this. That crow told all the rest.

As crows are found in nearly all countries, and as their habits are similar, we will refer to the

CROWS OF CEYLON.—Mr. Holman states that in Ceylon these birds are so audacious, that when the natives are returning home with baskets of provisions on their heads they are often attacked by these voracious birds, who pounce upon the contents and devour them. They will plunder children, and even dogs.

THE DOG AND HIS BONE.—It is amusing to see the art they use to dispossess a dog of a bone. No sooner has the animal laid himself down to enjoy his meal than a predatory covey descend and hover over him. One, more daring than the rest, alights, and advances towards him with the self-possession of an invited guest, when the dog lets fall his bone, and makes an indignant snap at the pertinacious intruder, which dexterously eludes the bite; while, at the instant the dog's attention is diverted, another crow, who has been vigilantly watching the opportunity, seizes the coveted treasure and bears it off.

AN UNLOOKED-FOR EXPOSURE.—Some years ago there was a tame crow about the Manse of Hay, in Orkney, which became so familiar as to visit all the apartments of the house, whenever doors or windows were left open. Crows, as is well known, are greatly addicted to pilfering, and our hero, not to be unlike his neighbours, perpetrated several achievements of this kind, of which the following is one of the more remarkable. One day about noon, a pack of cards was left on a table in the dining-room. The minister, who happened to go out with a friend, on his return wished to lay up the cards, but could not find them in any part of the house. Some time after, one of the family looking out of the window, which had been left open, saw all the cards arranged, face upward, on a garden wall which extended in a right line from the window. The mysterious disappearance of the cards was soon explained. The crow, strolling into the room, and seeing the unseemly sight of cards on a minister's table, had resolved to remove them ; and to give as much publicity as possible to the affair, laid them out one by one on the top of a wall. But little perhaps did he imagine that his exploit would one day shine in the annals of science, affording testimony to the sagacity of the crows of Orkney, and the unorthodoxy of its ministers.

CROWS IN CONCLAVE.—A lady resident at Dorking, in Surrey, was very much surprised some months ago at seeing about twenty crows sitting on a neighbouring tree, apparently in solemn conclave. After an hour's deliberation they all flew away, and shortly afterwards a crow, almost pecked to death, was found at the foot of the tree. Stanley mentions the same thing as occurring in Scotland and the Faroe Islands.

CROWS 'THE BLESSING OF GOD.'—Crows are useful in destroying caterpillars, which are often very destructive to wheat crops. In Barbadoes the negroes call the common black crow 'the blessing of God,' because it destroys the cockroaches which infest that place. Some years ago, in certain parts of America, the crows were driven away by incessant firing, through which the number of obnoxious insects so increased that the farmers found to their sorrow they had made a mistake by driving these scavengers away, and so they all agreed to stop the firing, and to suffer the persecuted crows to return and occupy again their old quarters.

JACKDAWS are also birds in black, and are common in England ; they live in flocks, and build their nests sometimes in trees, but more generally in old ruins, towers, and church steeples, where, dressed in their coats of glossy black, they attend, if not to the sacred duties of a divine service, yet to the parental obligations and pleasures of rearing their young

broods.   Could jackdaws learn the moral lesson which close
proximity to a church steeple, and the services which are conducted in the nave below, should teach to all mankind, they
would no doubt become less inclined to appropriate to themselves what is really not legally their own, but the property of
other beings.

It is worthy of notice that the same jackdaws will year after
year hover round, perch upon, and make their homes in the
towers of our ancient cathedrals and churches.   Why they
should do so may be difficult to explain, only that it is their
instinct to do so.   Are they lovers of what is antiquated?
Do they find a charm in and admire the different orders of
architecture in those fine buildings?   Or do they love the
chiming bells, the sound of which, mingling with their own
'Yak-yak,' they consider to be the sweetest of all music?   Our
poet Cowper, in playful humour, says of the jackdaw that
he is—

> 'A great frequenter of the church,
> Where, bishop-like, he finds a perch,
> And dormitory too.'

A SCHOLASTIC JACKDAW.—'Among my dumb companions,'
says W. F., 'my chief favourite has long been, and still is, a
tame jackdaw.   Every morning he makes his way to the
school-room, where he remains until the dismissal of the
school; he then leaves, and struts complacently forth among
the boys, with whom he is a great favourite, although at times
he is very mischievous.

'I may mention that though liberated on Sundays as on
other days, he never attempts to make his way to the room,
which is unoccupied on that day.   During the vacations, the
poor fellow approaches the school-door once every day at the
usual time, gives it a melancholy peck or two, and then
retreats to the barn, where he mopes most of his time away
till the return to school of his *"dear* boys."'

THE JACKDAW'S CARTE-DE-VISITE.—The following *carte-de-
visite* is given by the author of 'British Birds.'   He says:
'The jackdaw is a remarkably active, pert, and talkative
fellow, ever cheerful, always on the alert, and ready either for
business or frolic.   He is more agreeable than the raven, and
withal extremely fond of society; for, not content with having

a flock of his own folk about him, he often thrusts himself into a gang of rooks, and in winter sometimes takes up his entire abode with them. The rooks make him welcome. How do we know what amusement they (with their stolid gravity and solemn dignity) find in him, with all his fun and loquacity?'

Jackdaws, like rooks, are excellent weather prophets. If they fly back to their roosts in the forenoon, or early in the afternoon, a storm may be expected that evening, or early in the morning.

THE JACKDAW'S FAILINGS.—The Rev. J. G. Wood speaks of one jackdaw which had learnt the art of kindling lucifer matches, and thus became a very dangerous inmate, busying himself in this way when the family was in bed, though fortunately he seems to have done nothing worse than light the kitchen fire, which had been laid ready for kindling over-night. He frightened himself terribly at first by the explosion and the sulphur fumes, and burned himself into the bargain. But I do not find that, like the burnt child, he afterwards feared the fire, and so discontinued the dangerous trick.

Jackdaws are so mischievous that, after a gardener has planted cabbages or similar things, and has left his work, they will descend from their watching-place and pluck up every plant he has put in the ground. They will, when they have a chance, turn over and tear leaves out of a book, and unwind a ball of thread or worsted, etc.

DRUNKEN JACKDAW.—A curious story is told of a tame jackdaw which belonged to a publican living at Gilmerton, near Edinburgh. It appears that half a glass of whisky was left on the table, when Jackie flew up and began to drink the spirits, which seemed to please his palate so much that he drank a large quantity. He soon felt the effects—his wings dropped, and he half closed his eyes; he staggered in his walk, but managed to get to the edge of the table, from which he wanted to fly, but he dared not venture. At last his eyes closed, he fell on his back with his legs in the air, and looked altogether like death. He was then rolled in a piece of flannel and put in a box. Next morning the family expected to find him defunct, but it was not so. He had got out of the flannel, and when the door was opened, down he went into a court-yard, where he drank water very copiously from a basin, used by the fowls, several times during the day. Although he appeared to be no worse for getting drunk, he never again would taste whisky.

JACK AND THE OSTLER'S SON.—The son of the ostler at the Bush, at Staines, had a jackdaw, whose affection for him was the wonder of all who knew them. Such confidence had the owner in the jackdaw's devotion to him, that on one occasion when he was setting off from Staines to Hounslow, on horseback, he made a wager of two bowls of punch that the bird would obey his call and follow his route. He mounted, and then saying, 'Come, Jack; I'm going,' put his horse in motion. In a short time the bird's wings were extended, and he attended the progress and return of the ostler's son, leaving not the shadow of a plea for the non-payment of the bet.

If the jackdaw has some objectionable traits of character—and this cannot be denied—he has, nevertheless, some redeeming qualities. He is at all times sprightly, easily tamed, and learns to pronounce words and sentences very quickly, as well as to perform various tricks of the most amusing kinds. Although this bird has been unjustly accused of sucking the eggs of pigeons, doves, swallows, ducks, and fowls, he is vastly fond of peas and cherries. When these are out of season he visits gardens and fields for the purpose of feeding upon insects, of which he devours large numbers.

From what has been said in this chapter, it will be seen that our 'birds in black' are important and useful links in Nature's chain, and therefore, although sometimes destructive and mischievous in their habits, should not be wantonly destroyed.

> Having said thus much for our birds in black,
>   We hope you will give them their due ;
> Although they can't boast of much beauty or song,
>   They all have some duties to do.
> They toil hard and long for themselves and young,
>   Are loving, attentive, and true ;
> They stick to each other like Britons, I vow,
>   A hint both for me and for you.

CHAPTER VIII.

Whate'er the instincts of our birds may be,
  We have no right to view them with disdain ;
For e'en the eagle, owl, and heron too,
  Are useful links in Nature's endless chain.

S amongst human beings a wide difference exists in their power, status, and influence in society, so it is amongst the numerous members of the animal kingdom. Superiority and inferiority are inseparably connected with each other, while supremacy seems to be an indispensable necessity to the well-being, good order, and government of all living things.

The lion is usually denominated 'The King of the Forest.' Bees are controlled and, in a great measure, guided by the queen of the hive to which they belong. When the same sheep travel or pass through a gap in the hedge from one field to another, it is generally under the leadership of one of their number who may be more intelligent and daring than the rest.

Thomson says: 'Wild elephants march in troops, the oldest keeping foremost, and the next in age bringing up the rear, while the young and feeble occupy the middle ; but this order is not observed except in perilous marches. In ordinary cases the largest-tusked males put themselves at the head, and if they come to a river are the first to pass it.' In the feathered creation, the eagle is considered to be 'the king of birds.'

THE EAGLE.—This bird belongs to the order Accipitres, and family Falconidæ. There are many kinds of eagles, viz., the bald eagle, the black eagle, the white eagle, the sea eagle, and the golden eagle.

The structure of the eagle as a bird of prey is well adapted for rapine and tearing of flesh, its bill being very strong and curved at the top, and its talons sharp and powerful to hold its food securely. This bird is fierce and voracious, and seems to wage a war of extermination against all other birds and every animal it can overcome. It no doubt prefers the flesh of the animal it has killed, but will not despise that already dead. It has long sight and keen scent, which enable it to find its food with considerable facility.

The eagle is considered to be very courageous and bold, of great strength, and even to be magnanimous. Poets have called it the 'king of birds,' and in heathen mythology it is denominated the 'bird of Jove,' and regarded as the true emblem of might and dignity.

The golden eagle is considered to be the noblest of the family; often measuring in the expansion of its wings seven feet, and from the beak to the tail three feet. It is of a dark brown colour, beautifully shaded, and has been known to weigh sixteen pounds.

It has been frequently captured in the Highlands of Scotland, but it is becoming, like many more of our noble birds, exceedingly rare. Instances have been known of eagles attaining the age of 80 and 100 years, and a specimen is recorded to have lived at Vienna in confinement for 104 years.

One marvellous feature in the organism of the eagle is to be seen in its wings, which are very broad and concave in their under surface, for the purpose of enabling it to take good hold on the air, and to float in it with ease while its keen eye is in quest of food, which it may see afar off.

EAGLE's NEST.—The eagle builds its nest of large twigs, lined with layers of reeds, of a flat form, very strong, and several feet in breadth, and always in a situation that is dry and inaccessible to man, so that it may with its young be undisturbed. The same nest has been used for a succession of years.

Although the eagle is very rapacious it shows great tenderness and solicitude for its young, not only in procuring sustenance for them, but in teaching them to fly, and in watching their first feeble efforts to do so.

If the eaglets should falter for a moment and be in danger of falling, the mother at once darts underneath them, and catching them on her broad wings, bears them back in safety to their nest. This is a beautiful illustration of the fourth verse of the nineteenth chapter of Exodus, in which God speaks of His dealings with His people Israel, 'I have borne thee, saith the Lord, as on eagles' wings.' He had preserved them, saved them from danger, and brought them to a place of safety.

How Eagles Teach their Young to Fly. — This is beautifully described by Humphry Davy, who once had an opportunity of witnessing the proceedings of a pair of eagles after they had left their eyrie. He says : 'I once saw a very interesting sight above one of the crags of Ben Nevis, as I was going in the pursuit of black game. Two parent eagles were teaching their offspring, two young birds, the manœuvres of flight. They began by rising from the top of a mountain in the eye of the sun. It was about mid-day, and bright for this climate. They at first made small circles, and the young birds imitated them. They paused on their wings, waiting till they had made their first flight, and then took a second and larger gyration, always rising towards the sun, and enlarging their circle of flight, so as to make a gradually extending spiral. The young ones still slowly followed, apparently flying better as they mounted ; and they continued this sublime kind of exercise, always rising, till they became mere points in the air, and the young ones were lost, and afterwards their parents, to our aching sight.'

The above description is very illustrative of the tender care God has of His people, and of the methods He sometimes adopts to teach them useful lessons. This will appear especially so if considered in connection with what is recorded in Deut. xxxii. 11, 12, 'As an eagle stirreth up her nest, fluttereth over her young, beareth them on her wings, so the Lord did lead them.'

Strength of the Eagle.—Impelled by the demands of her own voracious appetite, and by maternal affection for her young, the mother eagle has often attacked not only small birds, but ducks, geese, and swans; even hares, fawns, and lambs have supplied this powerful bird of prey and her progeny with many a savoury meal; and instances are recorded of children having been seized and carried away for the same purpose.

The *Universe* contains the following story : ' The last known fact of this kind took place in the Valais in 1838. A little girl, five years old, called Marie Delex, was playing with one of her companions on a mossy slope of the mountain, when all at once an eagle swooped down upon her, and carried her away in spite of the cries and presence of her young friend. Some peasants, hearing the screams, hastened to the spot, but sought in vain for the child, for they found nothing but one of her shoes on the edge of a precipice. The child, however, was not carried to the eagle's nest, where only two eaglets were seen, surrounded by heaps of sheep and goats' bones. It was not till two months after this that a shepherd discovered the corpse of Marie Delex frightfully mutilated, and lying upon a rock half a league from where she had been borne off.'

EAGLE AND SALMON.—Eagles also love the delicacies of our rivers, in which they indulge whenever they find an opportunity of doing so. A salmon, or any other fish, is not, according to their ideas, to be despised. A story is told of a shepherd observing an eagle perched on a bank that over-hung a pool. The bird was intently watching the water; presently it darted down and seized a salmon. A terrible struggle ensued, and the water began to fly in all directions. When the man reached the spot he found both salmon and eagle under the water; the latter was not able to extricate itself, and the other was gasping for its life. So with one blow he broke the eagle's pinion and secured both the bird and its victim. The reason why birds of prey are drowned in this way is that in striking the fish with their talons they do so with very great force, driving them deep into the body of the fish, which suddenly plunges still deeper; and dragging the bird, who is not able to extricate itself, with it, keeps it under the water until life is extinct in both—the one dying from its wounds, the other for want of air.

A YOUNG GIRL AND HER EAGLE.—' Pliny relates that there was a wonderful example of the affection of an eagle at the city Sestos ; on which account that bird became afterwards much honoured in the neighbouring country. A young girl had brought up an eagle by hand. In return for this kind-ness, the bird would go in search of prey, and always returned with part of what it had procured to its nurse, to whom the eagle was devotedly attached. When the eagle grew stronger, she extended her depredations to wild beasts of the forest, and continually provided her mistress with stores of venison. At length the young woman took ill, and died, and when her funeral pile was burning, the eagle flew into the midst of it, and there was consumed to ashes with the corpse of the virgin. In memorial of this extraordinary event, the in-habitants of Sestos erected on the spot a stately monument, which they call Heroum, because the eagle is a bird conse-crated to God.'

To those who have but a superficial knowledge of animals the above statements may appear to be incredible ; but we may remark that by those who have closely noticed the great affection which animals often show to those who treat them kindly, the story may be believed as coming within the range of probability. What Sir Walter Scott said of dogs, that hardly anything could be told him about them that he could not credit, may be applied to other animals—not even except-ing the rapacious eagle.

A GOLDEN EAGLE AND A DECOY.—' Whilst staying a few days at Manhattan, a little town in Kansas, I spent some hours in the office of a dentist, Dr. C. Blackley, who is also an ornithologist, having stuffed a goodly number of the birds of the State. He was then occupied with a fine specimen of the common pelican (*Pelicanus communis*), one of a flock of over a thousand that passed over the town in the month of April, some of them alighting in the neighbouring marshes These birds are not unfrequent visitors to these far inland regions, and I have known them shot and brought to me from the alkali lakes in Colorado, both regions from 600 to 800 miles from the sea.

'The doctor told me an amusing incident of a day's wild-goose shooting in the vicinity. He took with him to one of the ponds frequented by wild geese, a stuffed specimen of the Canadian goose to act as a decoy. Having firmly planted his bird in the sand, with its wooden platform well covered over,

8

he lay behind the bushes awaiting a shot. Suddenly there was a rush of wings, and, like a flash of lightning, a golden eagle swept down on the decoy, knocking the bird over, and tearing out some of the stuffing. The eagle then sat down near his prey, staring with amazement at its remarkably quiescent character, as well as at the strange wooden appendage attached to its claws. Deeming there was something uncanny about such a goose, and that there might be danger in the neighbourhood, he prudently flew away.'

Although birds of prey are not eaten by human beings—at least not in England—they are of great service, especially in hot countries, where these rapacious scavengers devour offal and flesh, which might become putrid and produce pestilential fevers or contagious diseases.

In our own country they are no doubt useful in destroying vermin of various kinds. It is a great mistake to try to exterminate them, which we fear will be the case unless some check be placed upon the wanton and destructive proclivities of those who are ambitious to be considered 'good shots.'

It is only for a rare bird to make its appearance, and to come within gun-shot, to be the marked victim of some inconsiderate and ignorant man or boy who will try his best to deprive such a winged visitor of its life; and then, should he succeed in doing so, to carry it home in triumph as a trophy of cleverness in the use of the gun.

Kingfisher.

The noble eagle forms no exception to this foolish practice. Within the last few weeks a bird of this tribe was seen flying not far from Brighton, and was shot by some man who happened to be out with his gun in search of other birds.

THE KINGFISHER.— Of this bird there are upwards of fifty species, most of which are natives of Asia and Africa. There are some European varieties, which of course include the one found in our own country. Our English kingfisher has very beautiful plumage : colours of nearly all kinds adorn its body, giving it a gay and brilliant appearance. It has been very appropriately named, because it feeds almost entirely on fish, which it procures in streams, canals, and rivers. Its peculiar structure is remarkably well adapted for securing its prey. Its bill is long, strong, and sharp, and with it the bird can transfix a fish as with a spear.

The feet of the kingfisher are small, fitted neither for wading, standing, nor running, but perfectly suitable for perching on small twigs which overhang the water, and on which it watches and waits very patiently with its long beak directed downwards. The moment a small fish appears, the bird plunges into the water, and soon rises to the surface with it between its bills. It then perches on a tree ; and, after grasping the fish firmly by the tail, beats its head against a branch to deprive it of what life remains, and then swallows it, or conveys it to its nest to feed the young, which are very voracious, and always ready to devour the meal the parent has provided for them.

Baird informs us that ' this bird was known to the ancients by the name of " Halcyon," and many fabulous stories are told of it by the early writers. They supposed that it built its nest upon the surface of the sea, amongst the foam of the waves, and that it had the power of calming the troubled deep during the time of incubation. They only sat on their floating nest a few days, and during that short period, which was in the depth of winter, the mariner might, they said, sail in perfect security. Those days were hence termed " Halcyon days." The Tartars and Ostiacs—amongst whom this bird, or a nearly allied species, is found—preserve its skin about their persons in a purse, and reckon it a preservative against every ill. The feathers are used by them as love amulets, and they believe that if a woman is touched by a feather which floats on water, she will be induced to fall in love with the person who uses it.'

The kingfisher makes a home in the banks of rivers. During the time of incubation the male bird is very affectionate, working hard to supply the mother with food.

THE LAPWING.—This bird, sometimes called peewit, belongs

to the order Grallæ, and is found in every part of our own country, and especially in Scotland. It frequents open sea-shores and wide moorland wastes. It has a peculiar cry, which seems to be in harmony with the solitude of those places.

Lapwing.

Its plumage is very handsome; and on its head is a long crest lying backwards, which it can erect and lower at pleasure.

Although living in the midst of dreary wastes and by the lone sea-shore, it is a very lively, interesting, and frolic-some bird, and shows very strong maternal instinct for its young, which are reared in nests scooped out of heathy hillocks.

It is said the mother lapwing will, when her nest is likely to be discovered, tumble over and over as if it could not fly, or feign lameness to lure the sportsman away, which it usually succeeds in doing. In the month of October lapwings are considered excellent eating. Their eggs, which are sold at a very high price, are reckoned to be great delicacies, and are known as 'plover's eggs.' The plover belongs to the same family.

As the lapwing is a bird well known in our own country, we venture to give the following curious and somewhat romantic information on the authority of an English translation of the Koran, by G. Sale, p. 283:

SOLOMON AND THE LAPWING.—'And Solomon was David's heir, and he said, O man, we have been taught the speech of birds, and have had all things bestowed on us; this is manifest excellence. And his armies were gathered together unto Solomon, consisting of genii and men and birds; . . . and Solomon viewed the birds, and said, What is the reason that I see not the lapwing? Is she absent? Verily I will chastise her with a severe chastisement, or I will put her to death, unless she bring me a just excuse.'

Commenting on the above statements, the Arab historians tell us 'that Solomon, having finished the temple of Jerusalem, went in pilgrimage to Mecca, where having stayed as long as he pleased, he proceeded towards Yaman ; and leaving Mecca in the morning, he arrived by noon at Sanaa, and, being extremely delighted with the country, rested there ; but wanting water to make the ablution, he looked among the birds for the lapwing, called by the Arabs " Al Hesdbud," whose business it was to find it ; for it is pretended she was sagacious or sharpsighted enough to discover water underground, which the devils used to draw, after she had marked the place by digging with her bill,' etc.

Referring to the chastisement threatened by Solomon to be given to the lapwing, the above historians state it was to consist in 'plucking off her feathers, and setting her in the sun to be tormented by the insects, or by shutting her up in a cage,' which, if the reader will make himself acquainted with the cause of the lapwing's absence, as stated in the Koran, he will see would not only have been too severe, but that even the threat was altogether undeserved.

OWLS.—Owls are nocturnal birds belonging to the order Accipitres, and are very rapacious. They live upon insects, birds, reptiles, and small mammalia, which they hunt for after nightfall, when they are less likely to be noticed or disturbed. They love seclusion, and during the daytime hide themselves in thick ivy, deep forests, old ruins, or the fissures of rocks. Their eggs are white, and the young are covered with soft down.

The antipathies of individuals and of nations sometimes take very strange directions, and often exist against

Owl.

certain animals without their being able to give any really satisfactory reason for those antipathies. In every age and in every country superstitious notions have prevailed respecting owls, as they have been regarded with grave suspicion by some, and with horror as birds of ill-omen by others.

The Red Americans of the far West, although well acquainted with the habits and uses of the birds of their wide wild forests and prairies, look upon the owl as foreboding some disaster to themselves, and therefore give it no welcome. This is all the more remarkable because this bird is, in the Scriptures, always associated with desolation.

On the other hand, we find both the Greeks and Romans of ancient times always considered the owl as an emblem of wisdom, and sacred to Minerva. In some of the remote and less frequented districts of England, the owl's scream is regarded by the peasantry as ominous of ill-luck, sickness, or death in their own family, or in their circle of friends.

THE TAWNY OWL.—The screech of the tawny owl is certainly by no means melodious, nor calculated to inspire the timid and fearful with courage, especially if it should proceed from some ivy-mantled church, the ruins of some haunted castle, or thick dark forest during the

> ' Very witching time of night,
> When churchyards yawn, and hell itself
> Breathes out contagion to this world.'

And perhaps equally solemn and objectionable is its 'To-whit—too-whoo,' uttered as it passes a lonely cottage just at the very moment when the inmates are falling into a state of sweet and slumbering forgetfulness, or it may be as Morpheus is conveying them into dreamland, there to enjoy ambrosial sweets in Elysian fields.

The screech of the owl at such times may be very annoying and particularly unmusical, but what it has to do with human sickness, ill-fortune, or death, or why it should be considered ominous of these things, it is difficult to say.

THE OWL AND THE RAILWAY PORTER.—A railway porter belonging to an intermediate station in Sussex was sent, late in the evening, with a parcel to a gentleman's house some distance off. On his return it was dark, and having to pass through a wood, he took the wrong beaten path, which not only, detained him, but so perplexed and troubled him that he knew not how to extricate himself from his sad dilemma. While pondering over what he should do, he was startled by a sonorous 'To-whit—to-whoo!' uttered just over his head. 'Oh dear!' he said, 'who am I? Why, I am Jemmy P——, porter at the station, and I've lost my way. Oh dear, oh dear! do help me out, and tell me the right road.' 'To-whit—to-whoo!' again fell upon his ears; when poor Jemmy, now confused as well as perplexed, repeated : ' I've told 'ee. I'm porter at the railway station, and I'm lost, lost, lost !' The terrified man got no answer but a third ' To-whit—to-whoo !' which nearly drove him to desperation. What his state of mind might have become had he not heard the fluttering of wings amongst the trees, which convinced him that he had been answering the peculiar cry of some owl, it is very difficult to tell. He had mistaken the 'To whit—to-whoo!' for the question, 'Who are you?' hence the reason why he made a revelation of his status at the station, so promptly gave his name, and so beseechingly implored assistance in his dark dilemma. We have been told that as long as Jemmy remained at the station he was constantly teazed about this little adventure. He was therefore compelled, in order to avoid the annoyance, to take his departure to some other scene of employment.

The structure of the owl is deeply interesting. Being a bird of the night, it has a very acute sense of smelling, which helps considerably in guiding it to its prey. Like most other animals that are nocturnal in their habits, the owl has large and beautiful eyes, which are a necessity, because large eyes can take in a wider range of view, and admit more light than small ones. The eyes of fish that live deep in the water are larger than those that live near the surface.

For the uses and other particulars respecting the owl we refer the reader to our volume entitled ' Facts and Phases of Animal Life.'

AN OWL FEIGNING DEATH.—The following amusing story appears in the *Literary Miscellany :* 'Mr. Wales, of Bellingham, Massachusetts, relates a cunning trick of an owl caught poaching upon his premises. It entered a pigeon-roost and commenced killing right and left. The outcry of the victims arrested attention ; and on looking in, Mr. Owl stood motionless, like a sentry on guard. Mr. Wales took hold of him, but he did not stir. He carried him to the house, the bird being as rigid as if dead. He was laid on his back on the table, but there was no movement. As the family stood looking at him, he opened his big eyes, then turned upon his legs, and was at once wide awake. Mr. Wales said he feigned death, and did it to perfection, until convinced that he was out of danger—more ingenious than prudent.'

THE MINISTERS AND THE OWL.—It is said that two celebrated ministers, respectively named *Jay* and *Fuller*, were taking a drive in the country when a bird flew across the road. ' What bird was that—was it not a jay ?' asked Fuller.   ' Oh no,' said Jay. ' It is *fuller* in the eyes, *fuller* in the face, *fuller* in the breast, *fuller* in the feathers, and *fuller* all over.'   The bird referred to was a tawny owl.

How Owls are Hatched.—G. Manville Fenn says : ' It is commonly known that owls have two or three sets of young in the course of a season ; but, as far as I can make out, after sitting upon the first egg or pair of eggs, and hatching the birds, no further effort in incubation is made.   Directly after the owlets are out of the shell the hen-bird lays one or two more beautiful white eggs, but does not sit, devoting herself to feeding the insatiable little monsters she had started into life, and the warmth of their bodies hatches the next owlet. This one hatched, another egg is laid with the same result, that it is vivified by the young ones' warmth, escapes from the shell, and once more an egg or two occupy the nest, so that in the same corner in a shallow downy spot may be seen an owlet three-parts grown, another half-grown, another a few hours old, and a couple of eggs—four stages in all ; and, if inspected by day, the three youngsters will be seen huddled together in very good fellowship, one and all fast asleep, and the eggs in the coldest place outside.   The sight is not pleasing, as may be supposed from the above description of the young owls ; but if the eye be offended, what is to be said of the nose ?   Take something in a bad state of putrefaction and arithmetically square it ; the result will be an approach to the foul odour of a nest of owls in hot weather.   The reason is not far to seek, when it is borne in mind that the owl is a bird of prey ; but all the same, I have visited the nest earlier in the season, and found the place quite scentless, and that too at a time when ranged in a semicircle about the young were no less than twenty-two young rats and full-grown mice, so fresh that they must have been caught during the preceding night, the larder being supplemented by a couple of young rabbits. If, then, a pair of owls provide so many specimens of mischievous vermin in a night, they certainly earn the title of friends of man.   It may be argued that, inhabiting a pigeon-cote, the youngsters were the offspring of two or three pairs ; but as far as I can make out, a single pair only occupy the

cote from year to year, the young birds seeking a home elsewhere; and I may say for certain that the old birds do not
come near their young and eggs by day, generally passing the
time in some ivy-shaded tree while the sun is above the
horizon, far away from the cote containing their sleeping
babes. When fully fledged and nearly ready to fly, if the
strong scent is risked and a visit paid, the birds start into
something approaching to wakefulness, and, huddling up together, will stare and hiss at the intruder, ready to resist handling with beak and claw—and a clutch from a full-grown owl's
set of claws is no light matter; for Nature has endowed them
with most powerful muscles, and an adaptability for their use
that is most striking. When hunting for food, the owl glides
along on silent wing beside some barn or stack, and woe be-
tide the cowering mouse or ratling that is busy on the grain!
As the owl passes over, down goes one leg, and four sharp
claws have snatched the little quadruped from the ground,
the four points seeming to slope towards a common centre, so
that escape is impossible. Every seizure is performed with
the claws; the beak being reserved for dividing the animal
when too large, and not degraded into forming an instrument
for seizure or carriage of the prey. I have had owls calmly
seated upon my hand, but for a very short time; and I cannot
recommend ladies to try them for pets, for the sooner they
are perched elsewhere the more pleasant it is for the skin,
their claws being exquisitely sharp.'

AFFECTION AND GOOD SENSE OF AN OWL.—'A brown owl
had long been in the occupation of a convenient hole in a
hollow tree, and in it for several years had rejoiced over its
progeny, with hope of the pleasure to be enjoyed in excursions
of hunting in their company; but through the persecutions of
some persons on the farm, who had watched the bird's proceedings, this hope had been repeatedly disappointed by the
plunder of the nest at the time when the young ones were
ready for flight. On the last occasion an individual was ascending to their retreat, to repeat the robbery, when the parent
bird, aware of the danger, grasped her only young one in her
claws and bore it away; and never more was the nest placed
in the same situation.'

HERONS.—These birds belong to a family of the order
Grallæ, or wading birds. They are distinguished by long

slender necks, hard bills, short tails, and very sharp claws. They are formed for wading, frequenting marshy places, rivers, and lakes in search of food, which mainly consists of fish, but they do not object to reptiles, nor even to small mammalia. The heron family consists of numerous species, some of which are found in the North of Europe and Asia, and even in India, Syria, and Egypt.

The heron to which we refer is a permanent inhabitant of our own country and Scotland. It is of solitary habits, and save in the breeding season two of them are seldom seen together. When the water is low in the Severn river, herons may be seen here

Heron.

and there standing by the edge of the receding water, intently watching for some unlucky fish, which, as soon as it appears is pounced upon, dragged out of the water, and then has to take an aërial voyage a considerable height and distance in the bill or pouch of its captor.

The appearance of the heron just before and after it has seized its prey presents a very wide contrast. While gazing into the water it looks more like a motionless post just standing out of the sand-bed, than a living creature in search of food. No sooner, however, has it secured its victim than it stretches its long neck, expands its broad and hollow wings, and by the strong resisting power with which they are provided, the heron sails away, though heavily, with its living load, to some far-off place it has selected as its residence, or to its young family that may be anxiously awaiting its return to their nest.   •

The instinct of the heron teaches it to attend to the duty of seeking food by fishing just at the time and under circumstances the most suitable and favourable for that object. This bird may be seen watching in the water or on the shores and shallows when the fish come to them in search of insects, or when the fish are actively disporting in their native element. Cloudy days appear to be the most favourable to the fishing pursuits of the heron, simply because its body does not cast such a strong shadow in the water as it would do if the sun was shining brightly, and therefore the fish are less likely to be disturbed by the motion of the heron.

THE VORACITY OF THE HERON is very great, as this bird has been known to swallow several carp at a time, and to digest the whole in a very few hours, and then to go again after more food. It has also been known to catch up an eel, to hold it by the middle of the body, and then to fly away with it; not, however, without considerable inconvenience, as the twisting of the body of the eel this way and then that way very materially retards the flight of the bird.

THE HERON AND SPANIEL.—When attacked the heron will offer a fierce and determined resistance. 'A gentleman in Bothwell fired at and wounded a heron, and then sent his spaniel to fetch it out of the stream. As the dog approached the heron drew back its head, and then with all the force it could command stuck its beak into the dog's rib. The gentleman fired again and killed the heron, but had the mortification of seeing both the bird and the dog floating dead together down the waterfall.'

As before observed, herons will convey the fish they have captured many miles to their nests, as plaice and other fish, several inches long, have been found under the high trees in

Hobby.

which these birds build. Heronries were at one time common in England, but since heron-hawking has been discontinued they have become very scarce. A few, however, remain as mementoes of times gone by. Instances are on record of herons and rooks building their nests very near each other, and living, as near neighbours should live, on good terms one with another.

HAWKS.—Of these there are many kinds widely distributed, especially in northern latitudes. We will mention the *Hobby*. This hawk is about twelve inches long, and has a prominent hooked bill. It was formerly used in falconry, in catching larks and other small birds. The *Goshawk* is about twenty inches long. It is not much seen in England, being chiefly restricted to the Highlands of Scotland. It is not satisfied to prey only on small birds, but attacks even hares, squirrels, mice, and the larger ground birds. It is not a favourite of the gamekeeper,

Goshawk.

as it is very destructive to game of all kinds. The *Osprey*, or Fish-Hawk, is found in every part of Europe. It is nearly two feet in length, and lives principally on fish, on which it darts with great velocity and unerring aim. This bird builds its nest sometimes among rocks and in fir trees, and

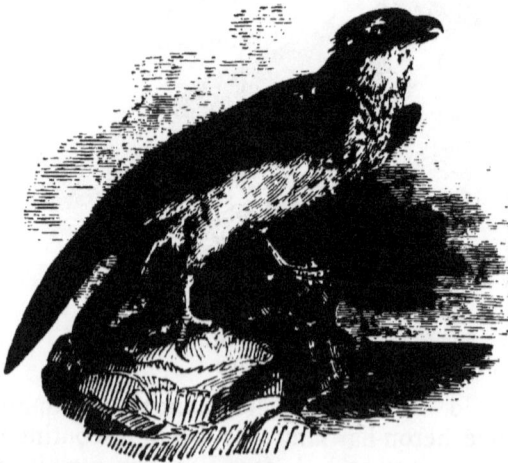

Osprey.

lays three or four eggs a little smaller than those laid by the common fowl. It appears that both the male and female birds attend to the duties of incubation alternately, and each in turn goes in quest of food.

CHAPTER IX.

PEACEFUL MONARCHS OF THE LAKE, ETC.

On lakes and ponds may oft be seen
  The swan, majestic as a queen ;
While ducks and geese, like maids of honour,
  Are near, as if to wait upon her ;
They glide with ease, now here, now there,
All well-built lifeboats, I declare.

THE habit which some men have of expressing their opinions of others by comparing their good or bad qualities with those of different animals is by no means uncommon. While this may be justifiable in some cases, it is not so in all. In many instances these comparisons are not only 'odious,' but, we think, a libel upon the animals whose names they use. ·Why should a man be called 'a drunken *dog ?*' or, as 'senseless as an ass ?' Dogs are not in the habit of getting drunk ; neither is the ass naturally senseless or void of understanding, but is as intellectual as many other animals which are credited with great sagacity.

When a man is told he is 'as silly as a goose,' it is intended to imply that he is considered to be defective in common-sense and in power of intellect, or that he has acted indiscreetly in something or other. We do not pretend to know everything about geese, but we confess we are at a loss to understand in what respects the 'goose is silly,' unless it is so for the reason once assigned to us by an old man, who said 'the goose was a silly bird because it was too much for one man's dinner, but not enough for two.'

If nothing stronger than this can be adduced in justification of the comparison, then it is evident the goose is libelled when-ever the expression referred to is used.

In this chapter we shall refer to three kinds of birds only, which may be included in those that are domesticated and belonging to our farmyards and homesteads. The first we shall refer to is

THE SWAN.—Swans belong to the order of birds in the system of Linnæus named *Anseres*, thus named from *Anser*, the type. It in-cludes all the web-footed water-fowls.

Tame, or, as they are sometimes called, mute swans, though well known in Eng-land, are by no means common. They may be seen in some parts of the different rivers of this country, on the lakes of some of our large estates, and occasionally in the ponds of our farmsteads. The male swan is called a *cob*, the female a *pen*, and the young *cygnets*.

Swan.

We remember seeing, some years since, about a dozen swans in a line, one behind the other, flying across one of the marshes in Gloucestershire at no considerable elevation. It was in the depth of a severe winter, and the birds were making southwards.

A promised visit 'to see the swans' constitutes a very great prospective pleasure to 'little folks,' and perhaps not less so to nursemaids, who in bright warm weather may be seen watching them glide so gracefully on the ornamental waters of our London parks.

Although in its general form the swan resembles the goose, it is, as seen upon the water, much more majestic in appear-ance than the goose, which may be owing to its greater size and its long and beautifully arched neck. This neck is nearly the length of the body, and consists of twenty-three vertebræ. The eyes are small, and placed very near the beak. The legs are short and bare, and the feet broadly webbed. It is in all respects well formed for the element in which it delights to live. On land the swan is stupid and awkward. It is noted for mischief and cruelty, especially in pairing-time, when the males fight with terrible fury, and the females, when attending their young, will attack anyone that approaches them.

FLIGHT OF THE SWAN.—'Although these birds fly heavily, they have been known, in a favourable gale, to fly 100 miles an hour. Some of them have lived fifty and seventy years. They subsist on aquatic insects, frogs, leeches, roots, leaves, and seeds, and even small fish. They are, when young, delicate eating, and if cooked with proper gravy, taste like goose and hare, or something between the two.'

Being aquatic birds, their nests, which are large and composed of grass, herbs, etc., are built close to their native element—that, is on the banks of the rivers, lakes, or ponds where they abide. They seldom lay more than six or eight eggs, which they have to sit upon about six weeks.

It is said that they are good prognosticators of a dry or wet season. If heavy downfalls of rain are likely to take place, swans build their nests much higher up the bank than when the season is likely to be an ordinary or dry one.

THE WHISTLING SWAN.—'The wild swan, elk, hooper, or whistling swan is a winter visitor to the British islands, but its native haunts are the northern regions of Europe and Asia.'

Swans generally pair for life, their whole behaviour offering a beautiful example of conjugal fidelity. The two birds show the greatest affection for each other, always swimming in company, and caressing one another with their bills and necks in the most interesting manner ; and should either be attacked, the other will show fight in the most vigorous manner, though, of course, the male is the more powerful and courageous. The young when hatched are very thickly covered with down, and are generally taken to the water by the mother when only a day or two old. There they are watched over by both parents with the greatest care until grown enough to provide for themselves. Swans were formerly designated *royal birds.*

AN ENRAGED SWAN.—Referring again to the pugnacious disposition of swans, we have been informed by a friend that a large number of these birds were kept, some years ago, in a reservoir on an estate not far from Swansea in Wales, and that on one occasion they were visited by some young people, who fed them with biscuits, pieces of bun, etc. When these were exhausted, they injudiciously tantalized the swans by throwing small stones and sticks to them. At this treatment one of the male swans became so enraged that he came out of the water, attacked one of the females, and so severely injured her about the neck and head, that she had to be taken at once to the hospital, where she was under medical treatment about a month.

A large scar remained on her temple as a memento of her visit to the swans, and as a proof that these birds will not be trifled with with impunity.

THE SWAN AND FOX.—The following incident is related in ' The Parlour Menagerie ' as having occurred at Pensy, Buckinghamshire. A female swan, while in the act of sitting, observed a fox swimming towards her from the opposite shore. She instantly darted into the water, and having kept him at bay for a considerable time with her wings, at last succeeded in drowning him ; after which, in the sight of several persons, she returned in triumph.

According to reliable information, it appears that in 1841 there were 437 swans on the Thames ; 232 belonging to the Queen, 105 to the Company of the Dyers, and 100 to the Company of Vintners. They are marked by a particular cut or nick on the upper mandible ; and on the first Monday of every August these marks are reviewed by the *swan-uppers* or *swan-hoppers*, as they are generally called.

Although the practice mentioned is resorted to in order to know to whom the different swans belong, it can hardly be regarded as the most humane method that could be adopted. We believe that in some parts of the Continent rings, sufficiently large and properly adjusted on the necks of these birds, are used, without inflicting suffering, or putting the swans to any inconvenience. This plan seems to be much more humane than that of cutting or nicking.

A SWAN FEEDING HER YOUNG.—The Rev. F. O. Morris says: ' On the Thames last summer I was amused by watching an old swan feeding her young ones in what seemed to me a novel and ingenious manner. Sitting on the water with her breast against the bank, she gathered from it the grass as far over as she could reach, and then, turning round her long neck, threw it over her back to the cygnets, who seemed quite up to the manœuvre, and were waiting and scrambling for it in the water behind her. My attention was called to it by the fisherman who was with me, who, though he had lived all his life by the banks of the Thames, said he had never witnessed this before.'

THE DUCK.—The duck belongs to the order *Natatores,* sometimes called by the name 'Palmapedes;' the first being from the word *nato,* to swim, the latter from '*palma,* the flat front of the hand, and *pes,* a foot, implying that the birds are *palm,* or *web-footed,* the toes of the feet being connected by a web or membrane.'

Duck.

The tame or domesticated duck is very common in England. A country ramble would be deprived of much of its interest if ducks could not be seen floating on our streams, rivers, and ponds. Their 'quack, quack,' though not musical, is very suggestive of farmsteads, pure fresh air, verdant meadows, fruitful fields, and the calm quietude of a rustic life.

The duck is a very useful and valuable bird, contributing materially in various ways to the sustenance and comfort of man.

The duck is easily reared, being by no means an epicure. It works assiduously to obtain its food, which consists of the refuse of animal and vegetable substances, snails, worms, slugs, and it does not object to a small frog, nor even to insects of various kinds. The duck lays a great number of eggs, and the young ones are reared without much difficulty and are easily fattened. We are informed that on a farm near Jedburgh there was a duck which in the spring laid black eggs. As the season advanced, the blackness gradually went off, till at the end of autumn the eggs were whiter than those of an ordinary duck. This duck was rather beyond the usual size. On the same farm there was another duck which laid two eggs a day. This fact was proved by locking the bird up, when one egg was found early in the morning, and another in the evening.

The common wild duck, or mallard, is no doubt the original stock of our domesticated duck. It is supposed there was considerable difficulty in weaning it from its former wild, free, and half-savage way of life. Those eggs that were first taken from the wild duck and set under an adopted mother produced very shy birds, which seemed unsettled, much inclined to roam, and very impatient to enjoy the unrestrained liberty of their predecessors. Even now the domesticated duck exhibits a propensity to go off in search of larger pools and marshy places, where it can enjoy its life in following out more fully the roving instinct with which it is endowed.

The strong proclivities of ducks in favour of watery places are necessary to preserve their size, strength, and beauty, which they do with frequent divings, washings, and sportings in their native element, and by finding in these places the kind of food suitable to their nature.

Although they are amphibious animals, if they were kept always out of water they would deteriorate both in size and value, and would be destitute of that flavour they possess when reared in proper places.

Duck or mallard shooting was at one time a favourite pastime in England. This method of securing them has been superseded by what are called

DECOY POOLS.—One of these has been opened very near the river Severn, in Gloucestershire. It is about two or three acres in extent, and a few feet deep. On its banks and from its bottom grow, here and there, reeds and rushes, and long grass. It is entirely surrounded by trees, which afford a screen from the burning sun of summer and a shelter from the cold winds of winter. It is thoroughly secluded, and an almost death-like stillness reigns over it, save when it is broken by the hoarse quack of its feathery inhabitants. Running from each of the four corners of the pool is an inlet about nine feet wide and thirty or forty yards long. It is arched over by net-work about six or eight feet high at the entrance, but gradually lowering to the extremity. To this is attached a circular net running about nine feet farther on, and closing at the end.

By the side of the inlet are placed several screens of straw and woodwork, which are so arranged that anyone may watch the whole surface of the pool without being observed by the ducks swimming thereon. When the proper season for catching wild ducks has arrived, the tame or decoy ducks are kept in the pool, for the purpose of attracting thither their relatives the mallards. When this has been done, they are further secured and finally captured in the manner now to be described. The keeper of the pool has a dog which he has trained to move gently on the bank of the inlet to draw the attention of the wild birds, which he invariably succeeds in doing. As soon as they begin to move towards him the dog disappears by passing through an opening among the screens; he then runs a few yards forward, and again appears on the bank. This is done several times, until the mallards have passed the opening of the arch of netting, when the keeper, who has been whistling for and watching the birds, and directing the movements of the dog, suddenly appears, lets down a portion of the netting, and thus secures his victims, who are now imprisoned in the inlet. They are then driven on and forced into the circular net, the mouth of which is easily closed by a mechanical contrivance. From this they cannot escape unless by accident or at the will of the keeper. From fifty to a hundred ducks have been caught in this way at one time. The decoy-ducks are of course taken out of the net and returned to the pool, while the remainder, that are deemed worthy, are sent off to market, or are presented to the friends and acquaintances of the possessor of the decoy-pool.

As the duck has to find and select its food more by the sense of touch than by sight, its bill is admirably organized for that purpose. It is abundantly furnished with nerves, and with a very sensitive membrane, which enable the animal to distinguish what is eatable from the mud, small stones, or other substances with which its food may be mixed.

The power of instinct is shown by the duck at a very early period of its existence, even as it relates to the food it eats. It is said that if you throw to ducklings but a few hours old a number of young flies they will greedily gobble them up; but if you throw before them young bees or wasps they will not touch them, but turn away and leave them.

The duck is by no means deficient either in affection or sagacity. If her little brood, which on the smooth silvery surface of the lake or pond look like floating balls of gold, should be in danger from some wanton schoolboy, or the rapacious appetite of one of the feline race, how stoutly will the mother resist the intruder and tormentor, and what fluttering anxiety she will exhibit until the threatened danger is past!

A BRAVE DRAKE.—The *Children's Friend* gives a story to the following effect: 'Some years ago, on the breaking up of the ice in Regent's Park, a duck got its foot fast in one of the ice-cracks. It gave cries of distress, when, flying in haste to the rescue, a fine drake was seen, who saw the difficulty at once. He set to work by chipping with his beak the ice around the imprisoned leg. It was soon freed, and several loud "quacks," with wonderful wagging of tails and flapping of wings, announced the release of the prisoner. A large crowd of spectators joyfully united in their congratulations over Master Drake's gallant rescue of his no doubt much loved *duck.*'

WILD DUCK'S NEST.—'Wild ducks,' says Johnson, 'are very artful birds. They do not always build their nests close to the water, but often at a good distance from it; in which case the female will take the young in her beak, or between her legs, to the water. They have been known sometimes to lay their eggs in a tree in a deserted magpie or crow's nest; and an instance has likewise been recorded of one being found at Etchingham, in Sussex, sitting upon nine eggs, in an oak, at the height of twenty-five feet from the ground; the eggs were supported by some small twigs, laid crossways.'

The keeper of the decoy-pool before referred to has informed us that the mallards, or males of the wild-duck family, usually select two, three, or four females each as companions, over whom they keep special guard, and to whom they pay the most assiduous and affectionate attention, while, at the same time, the groups keep apart by occupying different parts of the pool. Should one of these ducks leave her 'lord' to join another group, the mallard to which she belongs both by selection and mutual consent will evince the greatest jealousy and anger, and will chastise the delinquent duck for leaving his protection, and for showing signs of preference for another mallard or for the ducks he may have already under his care. Sometimes severe encounters take place between rival mallards before the matter relating to truant ducks can be decided. The victorious mallard, however, usually takes the duck under his protection, and seems to claim her by right of conquest.

SAGACITY OF DUCKS.—A correspondent of the *Animal World* communicates the following: 'Having, with a friend, been an eye-witness, I can vouch for the authenticity of what doubtless to a great many would appear an almost incredulous tale. When walking yesterday through Kensington Gardens our attention was attracted to a knot of people round an old tree, of which only the trunk remained, boughless and leafless, having been sawn off at the top, probably fearing its fall. We were told a duck had built her nest on, or rather in, the top of this old tree. Now we know ducks seldom build at any great height from the ground, so we waited to see if such was really the case. The mother-duck was walking round about the tree in a state of great excitement, fearing lest her young should be taken from her, and yet being unable to get them out of the tree by herself, loudly quacking as if asking help and a solution of her difficult problem. We had not long to wait, for, to the surprise of all present, the heads of two little ducklings appeared from a hole in the side of the tree, and about a foot from the top. They answered feebly to their mother's call, and then, after some deliberation, one of them came to the front, looked down on the bystanders, spread its tiny pinions (not being more than a week old and covered only with down), and fell heavily into the ready hands stretched out to receive it. The second seemed somewhat horrified at his brother's quick descent, but not feeling inclined to be left alone in his glory, plucked up all his courage and followed the good example set him; when both were restored to the anxious and now delighted parent. Such an instance of affection in ducks is, I think, most striking. The mother heeded not the crowd, but continued her call, when the little ones too, fearing nothing, replied in the manner described. Happily for them, some one was nigh to catch them as they fell, or I sadly fear the poor mother would have seen them either crippled or killed; for the tree was some height from the ground. Poor thing! very likely she may have found it difficult to secure a quiet spot to build, in a place so public as the Kensington Gardens, where the small boys seem, I am sorry to say, only too pleased to chase the ducks whenever an opportunity occurs, and thus this strange home was found for her young, quite out of their reach or sight. Probably she wished them to join her in the water, to be admired by all, and begin their aquatic education. Her puzzle was how to get them there, and so she continued to quack until the attention of some passers-by was attracted to the spot. I wish that I could better depict the scene—the anxious mother and her fearless, obedient children. I assure you it was well worthy the pencil of some of our talented artists, who at different times have so ably portrayed the leading characteristics of the animal world, and endeared them if possible still more to us all.'

A lady who keeps ducks also gives the following instance of sagacity: 'Once, very late at night, just as we were passing through the hall to go upstairs, we heard a great noise of ducks. It so happened that my husband was carrying a large paraffin lamp, and no sooner did the strong light appear in the hall than the noise very much increased. I opened a door which led into the garden, and no sooner had I done so than a duck positively rushed into the hall close to my feet, quacking most vociferously, her eyes glaring, and she beating her wings and moving her head about in great agitation. My husband turned out, and went into the garden, carrying the large lamp, and immediately the duck followed him. Outside she was joined by the rest of the ducks, young and old; but the old drake was nowhere to be found. The next morning he was discovered amongst the bushes, alive, but somewhat injured, having evidently been caught by a dog, who had dropped him upon hearing the alarm and seeing the light approaching.'

THE GOOSE.—This tenant of the farmyard and pond is so common, and its value, in various ways, so well understood, that a description of it would seem to be hardly necessary. Nevertheless, it claims, as well as the swan and the duck, at least a little notice. The tame goose, as now bred in this country, appears to owe its origin to the *Anser ferus*, the wild or grey lag goose, which is perhaps the largest species of the family, measuring five feet in extent of wings. Wild geese are found in nearly every part of the world, but especially in Northern countries, on the lakes, swamps, and marshes of which they breed in immense numbers.

As an article of food our tame geese are much esteemed; their quills are still extensively used as pens for writing, and their feathers and down are of great importance, not only as luxuries but in a commercial point of view. Very large numbers are bred in England, and, as in the case of turkeys, the consumption of them from Michaelmas until after Christmas is enormous.

In the fens of Lincolnshire geese are kept in great numbers. Some owners of these

Goose.

birds keep as many as a thousand breeders, not only to furnish geese for eating, but mainly for their feathers, for which they are plucked four or five times a year, and for their quills once a year. It is recorded that in 1783 one drove of geese consisting of 9000 in number was driven through Chelmsford on its way from Suffolk to London. Railway facilities have, however, rendered these tedious and we should imagine painful journeys unnecessary, although we fear the reprehensible practice often resorted to of packing them so closely in crates renders their transit by rail not only not easy, but even painful. Such a practice we must condemn as exhibiting not only a parsimonious feeling, but a lack of common humanity. On the authority of 'All About Country Life' we are glad to state that the unnatural and cruel practice of plucking live geese is getting obsolete. We shall now describe at least some parts of the

STRUCTURE OF THE GOOSE, which are distinguishing features or characters of all the different kinds of geese, whether wild or tame. We are informed by 'Treasures of Natural History' that 'the bill is the first great distinction of the goose kind from all the feathered tribes. In all other birds it is round or wedge-shaped, or crooked at the end; in all the goose kind it is flat and broad, formed for the purpose of skimming ponds and lakes of the mantling weeds which grow on their surface. The bills of other birds are composed of a horny substance throughout, formed for piercing or tearing; but birds of this genus have their inoffensive beaks sheathed with a skin which entirely covers them, and are only adapted for shovelling up their food, which is chiefly confined to vegetable productions; for though they do not reject animal food when offered to them, they contentedly subsist on vegetable, and seldom seek any other.'

We may add that the long necks of geese are of great advantage to them when feeding in the way just described, because while remaining in one spot on the surface of the water they can throw out their long necks and so secure all the food that may happen to be within the radius of the half-circle they can so easily describe.

As feeders on vegetable productions, geese by no means ignore tender blades of grass, which they seem to devour with the greatest pleasure, and on which they thrive; as well as on the scattered grain they sometimes pick up in stubble-fields after the harvest is gathered in. In many cases, however, where large numbers are reared for the market they are fed systematically by prepared food of various kinds, in order to increase their weight and of course their value too.

It has been remarked that none of our domestic birds are so apt to bring forth monstrous productions as geese—a circumstance which has been attributed to the excessive fatness to which they are liable. The liver of a fat goose is often larger than all the other viscera, and was a dish in so great reputation among the epicures of Rome, that Pliny thought it deserved a serious discussion to whom the honour of inventing so excellent a dish was due.

The feet of the goose being thoroughly webbed, afford considerable facilities to it to pass easily and at considerable speed through the water. Although our tame geese are under no necessity of flying, and so have not to use their wings for that purpose, there is great strength and powers of endurance in them, especially in those of the wild kind, who often travel, at a considerable elevation, many miles in search of food, and return to their resting-places for the night.

In the autumnal season of the year we have often watched with boyish glee the flight of great numbers of geese, arranged one after the other in the shape of the letter V, flying across the country from the river Humber to the Yorkshire wolds,

It would appear that wild geese are aware of the necessity of using every precaution for their own safety while feeding in a stubble-field, or in any other place. We have been informed that, while they are thus engaged, one or two of their number are appointed to act as sentinels, whose duty it is to look out for an approaching enemy, whether man or quadruped, and at once, when seen, to sound the note of alarm. When this is heard, all the geese rise simultaneously to a great height, form themselves into two lines as described, then follow their leader, who constitutes the first link in this acute angle.

The tame goose lays from seven to twelve eggs, sometimes beginning in January, but more commonly in February.* The time of incubation lasts about thirty days. Having reared her brood, it often happens that the same goose will commence laying eggs again at harvest-time, and will sometimes produce as many in autumn as in the spring. In egg-laying, geese arrive at the greatest usefulness after they are three years of age. The longevity of the goose is proverbial. Some have been known to be thirty years old, and mention is made of one who reached threescore years and ten.

The custom of eating geese at Michaelmas, it is said, originated with Queen Elizabeth, who happened to be feeding on roast goose when she received tidings of the destruction of the Spanish Armada, and in commemoration thereof continued to dine on the same dish when the anniversary of the day recurred.

The *Evening Standard* says, ' A critical epoch in the life of the goose is the season which sets in on the 29th September. Queen Elizabeth used to be accredited with the origin of the custom of causing flocks of these noisy creatures—that " gabble o'er the pool," especially, according to Goldsmith, at that hour when the noisy children are "just let loose from school"—to be on that particular day served up to table and form the dish of honour, taking precedence of all other dishes. Recent researches, however, have traced the custom back to a date much more remote, and to countries on the Continent far

---

* It is said that if, after a goose has laid seven or eight eggs, one egg per day be taken from the nest, the goose will continue to lay until she has produced about thirty eggs, more or less. If none be taken from her, when she finds about twelve or thirteen eggs under her she ceases laying, and attends exclusively to incubation.

distant from our own.   In the England of our own day the
goose, as a rule, is reserved for the hospitable dinner-tables of
Christmas and thereabouts.   Then the stubble-fed bird is in
its primest condition, and its succulent flesh, when tempered
with apple sauce, is most palatable.'

A CARNIVOROUS GOOSE.—The following curious information appeared in *Nature*, of April 19, 1879:

' I enclose to you an account of a golden eagle, which I have
reason to know to be authentic.   The possibility of a bird so
purely graminivorous as a goose being taught to eat flesh,
and acquiring the power of digesting it, is extremely curious.
It is well known,' however, that cows are largely fed on fish
offal in Scandinavia, and I have heard of a Highland cow devouring a salmon which an unwary angler had hid among
fern on the banks of a river in Sutherland.

'ARGYLL.

' Isola Bella, Cannes, April 7th.'

' March, 1879 : There is in the possession of W. Pike, Esq.,
at Glendarary, in the island of Achil, co. Mayo, a golden
eagle, now about twenty-five years old, which was taken from
the nest and brought up in confinement.   This eagle, in the
spring of 1877, laid three eggs, which Mr. Pike took away,
replacing them with two goose eggs, upon which the eagle sat,
and in due time hatched two goslings.   One of these died,
and was torn up by the eagle to feed the survivor, who, to the
great tribulation of its foster-parent, refused to touch it, together with the other flesh with which the eagle tried to feed
it, Mr. Pike providing it with proper food.   The eagle, however, in course of time taught the goose to eat flesh, and (the
goose having free exit and ingress to the eagle's cage) always
calls it by a sharp bark whenever flesh is given to it, when
the goose hastens to the cage and greedily swallows all the
flesh, etc., which the eagle, tearing its prey to pieces, gives it.

' I saw them in May, 1878, when the goose, being a year
old, had made a nest in the eagle's cage, and laid eleven eggs,
and the two birds were sitting side by side on the nest.   I
hear from Mr. Pike that he did not allow them to hatch out,
fearing that it might interfere with their attachment to one
another.

'The eagle is very tame and fond of Mr. Pike; he goes
into the cage, and it allows him to handle it as he likes, but

will not allow anyone else near it. It never attempts to get out of the hole made for the goose to go in and out at.'

Geese feel so much concern for the safety of their young that the gander will not only hiss defiance and displeasure at an intruder, be it dog, pig, horse, or man, but has been known to attack children so savagely as to inflict upon them serious injury. These birds are, however, susceptible to kindness, as the following anecdote will show, which we copy from the 'Parlour Menagerie':

A GRATEFUL GANDER.—'An old gander of a surly temper, in the habit of attacking everyone that passed, chanced to wander up a narrow drain, from which he was unable to extricate himself. A labourer observed the awkward predicament into which the bird had got, and at once pulled him out. As if to show his gratitude, the gander was afterwards in the habit of following his deliverer like a dog, and allowing himself to be handled in any way the man chose. This freedom was confined solely to his deliverer—to all others he maintained his former pugnacity.'

'AS SILLY AS A GOOSE.'—This expression, as before stated, is often used to denote the want of good sense on the part of some one who may have performed some absurd or foolish act. This libel on the goose is, however, very clearly disproved by Mr. St. John, in 'Wild Sports of the Highlands.' He says: 'Even a tame goose shows much instinct and attachment; and were its habits more closely observed, the goose would be found to be by no means wanting in general cleverness. Its watchfulness at night-time is proverbial; and it certainly is endowed with a strong organ of self-preservation. You may drive over dog, cat, hen, or pig, but I defy you to drive over a tame goose. As for wild geese, I know of no animal, biped or quadruped, that is so difficult to deceive or approach. Their senses of hearing, seeing, and smelling are all extremely acute; independently of which, they appear to act in so organized and cautious a manner, when feeding or roosting, as to defy all danger. Many a time has my utmost caution been of no avail in attempting to approach these birds; either a careless step on a piece of gravel, or an eddy of wind, however light, or letting them perceive the slightest portion of my person, has rendered useless whole hours of manœuvring.'

CHAPTER X.

BIPED TENANTS OF THE FARMYARD.

One tenant of the yard and field
  Is often heard to bellow ;
One crows a cock-a-doodle-do,
  And one's a grunting fellow.
Pintados seem to say, "Come back ;"
  But turkeys only gobble;
The strutting peacock, finely drest,
  Screams out as if in trouble.

IF we were to form our opinion of the intrinsic value of everything by its external appearance, we should be very often in error. In nothing is this more clearly seen than in nature, animate and inanimate. The sunflower, towering six feet high, with its golden boss turned to the sun, looks beautiful and majestic, but it gives no fragrance grateful to the olfactory nerves. The violet, blooming on a bank by the wayside, and bending its head modestly downward, is often hidden from sight, but from its tiny leaves comes a sweetness which tells us at once it is there. It would be in vain to look for a combination of all excellences in any one created thing, but some may be found in all of them. Animals in their life and surroundings particularly show the truth of this statement.

If the peacock cannot sing, he shows in his variegated feathers, in the brilliant purple and green colours of his head and breast, the beauty of Nature, and the most delicate touches of her pencil. If the turkey can claim no ability as a feathered chorister, but on the other hand has a very unintelligible and disagreeable way of expressing himself, he is of

more use to man after death than would be the sweetest bird-
singer ever heard in our fields or woodlands. The pintado
cannot boast either of a commanding and graceful form of
body, or of splendid feathers; yet its eggs are good eating,
and its flesh, like that of the turkey, is esteemed a delicate
article of food. And so it is throughout animated nature:
every animal possesses some useful, valuable, and peculiar
property; what it may lack in one way is made up in another.

Having in a previous volume—'Facts and Phases of
Animal Life'—devoted a chapter to the common fowl family,
we now propose making a few remarks on the following
feathered tenants of the farmyard : namely, the peacock,
turkey, the Guinea-fowl, or pintado, and the pigeon.

THE PEACOCK.—This bird is, in many respects, the very
reverse of the owl. The latter avoids as much as possible all
contact with human beings, hides himself by day, and comes
stealthily out at nightfall to hunt for his living. He prefers
seclusion at all times, and, as we have shown, loves to linger
in old towers, ruins, and crevices of rocks. The peacock
lives in the society of man, and in the open light of day seeks
his food and attends to the duties his necessities impose upon
him. He appears to be fond of display, is very vain, and
loves to be admired. If you look over the wall of the farmyard you may see him quietly feeding with his family, while
his long tail or train lightly sweeps the ground on which he
walks. But the moment he is aware of your presence, he
greets you with a peculiar noise, spreads out his tail-coverts,
which are of the most dazzling colours : yellow gilded with
different shades, green running into blue and bright violet,
the centre being of a beautiful soft black, and all varying according to its different positions. The tail is of a brownish
colour, and is hidden underneath the tail-coverts, or those
long graceful feathers sometimes called 'the peacock's train,'
and which, when elevated and spread out, resembles a large,
gorgeous fan. Whether the spreading out of this 'feathery
fan' of the peacock depends more upon anger or pleasure it
may be difficult to determine. When not disturbed by in-
truders of any kind, the peacock may often be seen on an
eminence, a gate or wall, with his tail-feathers elevated and
opened out as wide as possible, as if to make his beauty all
the more conspicuous. The head of this bird is also of rare

Peacock.

beauty. It is adorned with feathers of a green colour, and bordered with rich gold. From the head to the breast there is a charming blue colour, mixed with green and gold. He has a stout beak, strong legs, and rounded wings. The female, or pea-hen, cannot boast of these rich colours, hers being alto-gether of a sombre hue.

The peacock has, however, his seasons of humiliation. When the leaf falls he loses his ' train of gorgeous feathers ;' and will then creep into corners or anywhere out of the sight of human beings until the return of spring, when he is fur-nished again with a feathery plume. His cry or voice is most discordant, and he is very voracious, which lessens his popu-larity.

PEACOCKS DEFENDING A YOUNG THRUSH.—' Two summers ago,' says a writer in the *Animal World*, 'I was staying in a country house in Scotland, the happy home of various pets. Among these was a fine cat called "Ben," and a pair of pea-cocks. The three were on friendly terms, and might often be . seen in the garden ; the peacocks pacing with measured steps, and the cat either scampering wildly up and down the trees, or pretending to doze on the lawn, till the sweeping train of the peacock came within reach, when a sudden clutch would prove pussy not so unconscious as he looked, and quite unable to resist a practical joke at the expense of his friend's dignity. The windows of my room looked out on the garden, and I often amused myself by watching Ben's sly pranks. One morning I was startled by the shrill note of a bird evidently in distress. On looking out, I saw the cat on the grassy bank crouching over a dark object. While I watched, the two pea-cocks hurried round the corner with ruffled feathers and out-stretched necks. They had evidently heard the harsh outcry of the captive, for to my surprise they both flew at the cat, and with a few vigorous pecks made him drop a bird, which hopped off into a bed of stocks; after which the peacocks marched away. For a minute the cat stood still, then crept round the flower-bed ; another minute, and a loud squeak announced the success of a rapid pounce. Again the pea-cocks rushed into view, and again rescued the bird, a fine young thrush. This time the cat was on the alert, and as the bird hopped slowly off he stole after. The peacock, observing this, swept quickly round, and facing the cat, put himself before the thrush, while his mate stood near. By this time I felt a strong interest, and called my husband from the next room to watch the end, ringing the bell at the same time for my maid. Certainly two or three minutes must have elapsed before she came upstairs—at my request went down, through the long drawing-rooms, and out of the glass door, and all this time the peacocks defended the thrush in the most spirited manner. The cat made most persistent efforts to creep round and seize the poor young bird, which was now standing bewildered and silent behind his defenders. The peacocks always fronted the foe, sweeping rapidly round as often as the cat tried to get behind, and they did not relax their vigilance till the maid arrived and carried off the cat. The thrush then hopped merrily away into the shrubbery, and the peacocks resumed their usual measured pace.'

A SINGULAR FANCY.—It is said the peacock is not fond of having his roost prescribed to him. A gentleman residing in the suburbs of Edinburgh had a peacock which uniformly went to roost at nightfall in the avenue of one of the public parks of the city, where it was liable to be stolen and was frequently annoyed. There were many large trees on the property of the gentleman, but the stupid bird persisted in visiting the avenue, where, as might have been expected, it ultimately became the prey of thieves.

Referring again to the pea-hen, we quote the following account of her nest as given by Brown. He says: 'In the nest of a pea-hen which we lately examined, we observed that the mother had taken care to choose a very sheltered spot, the nest being overhung by a low branch of a spruce fir, which was suspended over it like an umbrella, and completely protected it from rain and dew. Another circumstance was still more remarkable. It is well known that female birds, for the most part, wear off a considerable portion of the feathers from their breasts by their frequent movements in turning their eggs. Now, as her eggs were placed on the bare earth, no grass growing under the grip of the spruce branch, the breast of our pea-hen must soon have been rubbed bare of feathers. Foreseeing this event . . the careful creature prepared a soft cushion of dry grass upon which her breast might rest. This cushion was placed on the most exposed side of the nest, but no part of it under the eggs themselves.'

The foreign relations of our peacock are those of Japan, China, and Thibet. The painting of a Japan peacock was sent by the Emperor of the former country to the Pope, since which time it has been known in Europe. The date at which the present was made is not stated.

THE TURKEY : ITS ORIGIN, ETC.—This bird belongs to the order *Gallinæ*, is very common in England, and constitutes in some districts a very important article of commerce, especially in Norfolk and Suffolk, whence they are sent in large numbers to the London market. If we may judge from the immense quantity of these birds which may be seen in our shops about Christmas-tide, it is evident their flesh is highly esteemed as food, quite as much so as that of the goose, and in some cases even more so.

For some time an erroneous idea prevailed that this bird was a native of Turkey, which in all probability arose from the circumstance that about the time it was introduced into England we were importing from that country many different things which were regarded as great luxuries, and as the flesh of this bird was deemed to be one, it was classed with the others; and so, without troubling to find out where it really did come from, it was concluded that it was brought from Turkey, and on this account received the name by which it is known.

Baird states that the common turkey (*M. gallopavo*) is the type of the genus. Mr. Gould has lately shown that this bird is a native of Mexico, and is the true origin of our domestic species. It appears to have been first imported into Europe by the Spaniards in the year 1530. For many years, however, this species has been lost sight of; and another, closely allied, a native of the United States of North America, has been generally considered the origin of our well-known domestic bird. In their wild state, turkeys attain a large size, and birds of from twenty to thirty and even forty pounds' weight are often met with.

STRUCTURE OF TURKEYS.—Familiar as the personal appearance of turkeys may be to most people, we may notice that in a state of domestication they vary in colour: while some are grey and almost black, others are white, and black and white. When the males are about three years old they put out a tuft of hair which hangs from the breast. From the head and

Turkeys.

neck, and from underneath the bill, hangs a fleshy dilatable appendage, which, when the male bird is excited by fear, or agitated with desire, enlarges, and becomes alternately red, white, blue, and yellow.   When thus agitated the male also erects his tail, and spreads it like a fan, his wings droop and trail on the ground, and he struts about with a solemn pace, and assumes all the dignity of the most majestic of birds.

Although he may have a very high estimate of himself, he is little less than a 'coxcomb among birds,' and, with all his ostentatious display, whenever he essays to open his mouth he utters a kind of gurgle which is unmusical and disagreeable. The noise he makes may, however, not only please himself, but may possess a great charm and power for the female turkeys.

The male bird has a great antipathy to everything that is of a red colour, and has often been known to pursue children and others wearing red dresses, particularly when they have shown any fear of him.   When these birds are being driven to market, those who have the charge of them often carry a red flag on the end of a long stick, by means of which the drivers manage them with much greater facility than they would be able to do without it, because the hatred these birds have to a red colour urges them onward with greater speed.

AFFECTION OF TURKEY HENS.—'The turkey hen is a timid, inoffensive bird, and greatly attached to her young.'   Buffon says, 'When the hen turkey appears at the head of her young, she is sometimes heard to send forth a very mournful cry, the cause and intention of which are unknown; but the brood immediately squat under bushes, or whatever presents itself for their purpose, and entirely disappear; or if they have not a sufficient covering, they stretch themselves on the ground, and lie as if they were dead, in which state they continue perhaps a quarter of an hour, or longer.   In the meantime, the mother directs her view upwards with fear and confusion, and repeats the cry that laid her young prostrate.

'Those who observe the commotion of the bird and her anxious attention endeavour to trace the cause; which is always a bird of prey, floating in the clouds, and whose distance withdraws him from our view, but who cannot escape the vigilance or penetration of the active mother: this occasions her fears, and alarms the whole tribe.'   'I have seen' (says the Abbé de la Pluche) 'one of these creatures continue in this agitation, and her young in a manner riveted to the ground for an hour successively, while the bird whirled about, ascended, or darted down over their heads.   But if he at length disappear, the mother changes her note, and utters another cry that re-vives all her brood ; they run to her, flutter their wings, tender her their caresses, and undoubtedly relate all the dangers to which they have been exposed.'

REMARKABLE INCIDENT.—The Rev. H. J. Swale, of Ing-field, is in possession of a turkey cock, the female of which was recently set on a number of eggs.   Whilst the hen was so engaged, the cock, it appears, chanced to find a nest of fourteen duck eggs not far off, on which it forthwith com-menced to sit, and whilst the process of incubation was going on it is said to have been a faithful sitter.   It succeeded in bringing out four fine ducklings, and the remainder of the eggs were found to be rotten, which may be accounted for by their probable long exposure to the weather before they were found by the turkey.

A GALLANT TURKEY COCK.—In ' Anecdotes of the Animal Kingdom,' it is stated that 'a gentleman of New York re-ceived, from a friend at a distance, a turkey cock and hen and a pair of bantams, which he put into his yard with other poultry.   Some time after, as he was feeding them from the barn-door, a large hawk suddenly made a pitch at the bantam hen ; she immediately gave the alarm by a noise which is natural to her on such occasions ; when the turkey cock, who was at the distance of about two yards, and no doubt under-stood the intentions of the hawk, as well as the imminent danger of his old acquaintance and companion, flew at the marauder with such violence, and gave him so severe a stroke with his spurs, when about to seize his prey, as to knock him from the hen to a considerable distance ; and the timely aid of this faithful auxiliary completely saved the bantam from being devoured.'

A STRANGE THEFT.—'A jeweller of Manchester, being away from home for two days, left in his shop a tame turkey. This bird, urged by hunger, swallowed about £5,000 worth of cut diamonds, and flew through a window in search of more substantial nourishment.   Being caught, killed, and cut up by a cook, he strangely puzzled his new possessor.   But the honest man lodged the diamonds in the hands of his attorney; who restored them to the jeweller, when the news-papers made known the loss he had sustained, which was attributed to some very adroit thieves, as he never dreamt the turkey had been the depredator.'

THE GUINEA FOWL, OR PINTADO.—Guinea fowls, like turkeys, belong to the order *Gallinæ*, and are natives of Africa, but now domesticated in almost every part of Europe. Their bodies are round, tails pendent, necks and legs short, heads small, and their plumage a dark bluish-grey, sprinkled with

Guinea Fowl.

round white spots of different sizes. Their flesh is considered to be, in flavour, nearly as excellent as that of the pheasant : they lay many eggs, which are smaller than those of the common barn-door fowls, but richer in quality. In egg-laying, the guinea-hen exhibits great secretiveness as to the places in which she deposits her eggs ; and it often happens that she produces her young brood before the place where she has hatched them has been discovered. It is on this account, and the great difficulty in rearing them, that their numbers are so small, compared with those of other feathered tenants of the farmyard.

Although these birds in most cases retain much of their native wildness, and seem inclined to roam just as they list, yet if taken when young and properly trained, they become very tame.   Mr. Bruce informs us that when he was on the coast of Senegal, he received as a present from an African princess two guinea-fowls.   Both these birds were so familiar that they would approach the table, and eat out of a plate. When they had liberty to fly about the beach, they always returned to the ship when the dinner or supper-bell rang.

PIGEONS. — We give a place to pigeons in this chapter, principally because they constitute a connecting link between passerine birds and the poultry of our farmyards, and because many of them are really domesticated.   The pigeon family is a large one, comprising many varieties, found in nearly every part of the world, some even in the coldest latitudes.

'The pigeon family' (says Baird) 'may be divided into three groups or sub-families: *Treroninæ*, or tree-pigeons ; *Columbinæ*, the true pigeons ; and *Gourinæ*, the ground doves. Of these there are several species, but as they are too numerous to refer to in detail, we may state that the species of *Columbinæ* not only exist in great numbers, but almost everywhere.   This genus comprises the *cushat*, or *ring-dove*, which is the largest of our British pigeons, and builds its nest in ˌthe boughs of our trees.   The *stock-dove*, or *wood-pigeon*, lives in hollow places in decayed trees in Great Britain, and was at one time believed to be the

Wood Pigeon.

parent or stock of our domesticated pigeons—hence the name.

'The real parent, however, is now considered by ornithologists to be the *rock-pigeon* (*C. livia*) ; and from it are derived not only the common pigeon, or inhabitant of the dovecot, but all those numerous varieties of domesticated pigeons so highly prized and fostered with such care and attention by the amateur breeder or pigeon-fancier. . . .'

Group of Pigeons.

In the wild state, the *rock-pigeon* lives and breeds in holes
of rocks in this country, migrating southward in winter.

The principal food of pigeons is grain; they drink much,
not like other birds at intervals, but by a continued draught
like quadrupeds. They pair in the season of love, and pay
court to each other with their bills. The female lays two
eggs, from which usually come a male and female. The

Rock Pigeon.

parent birds act on the mutual or co-operative system, as
during the time of incubation they sit on the eggs by turns,
and share in the duty of feeding their young. The domestic
pigeons breed every month, and their fecundity is immense.
It has been calculated that if two pigeons were to hatch nine
times a year, in four years they would be the progenitors of
14,760 young.

While other kinds of birds may feed and rear those fledglings that have been deprived of their parents, it is said that no other birds can perform the same good offices to young pigeons, only pigeons themselves. During the first few days after the young are hatched they are fed by the mother with a substance resembling thick milk—a kind of food macerated by her own digestive organs. It has been observed that towards the latter part of the time of incubation the breast of the hen pigeon enlarges, and is exceedingly prominent immediately after her young are hatched. This is caused by the milk-like fluid referred to, without which young pigeons could not be reared. It would therefore appear that the once common practice of making April fools of simple folks by sending them for a pennyworth of ' pigeons' milk ' is based on fact.

FANCY PIGEONS.—Owing to cross-breeding, varieties of pigeons have, during the last few years, considerably increased in this and other countries. Some of these are known by certain peculiarities of structure, habits, and capabilities ; hence we have *tumblers, pouters, jacobins, fantails, owls, trumpeters, barbes, turbits, nuns, dragoons,* and *archangels.* Probably the most prominent and useful pigeon is the one known as the

CARRIER PIGEON, highly valued and immortalized by Anacreon and other poets, as the bearer of love-epistles, and by historians as the messenger sent by beleaguered hosts to friends at a distance. Of late years this kind of pigeon was employed both in England and on the Continent by those engaged on the racecourse or in the prize-ring, and in many stock-jobbing transactions. The use of the electric wire at the present time has rendered the services of these birds less necessary than formerly.

It is a matter of surprise that carrier-pigeons taken a long distance from home should, when let loose, so unerringly find their way back. They appear, like migratory birds, to possess either an additional sense or instinct of a very superior order. We have been informed by a gentleman much interested in pigeons, and particularly in carrier-pigeons, that the plan adopted by the latter in their flight homewards is as follows: first ascending and satisfying themselves as to the direction they must take for home, they describe a circle, in doing which their eyes take in the panorama of a large extent of

country; the next circle they describe is wider, and the next
still larger, and so on; the pigeons, however, moving all
the time in the direction of home.   Being, as most birds are,
long-sighted, they soon, from their great altitude, discover on
the landscape objects which are familiar to them; they then
take a direct course for the destination or home they wish to
reach, and which they usually do in an incredibly short space
of time.

Lithgow assures us that a carrier-pigeon would carry a

Carrier Pigeon.

letter from Babylon to Aleppo (which to a man is usually
a thirty days' journey) in forty-eight hours.   We have read
of a gentleman who, some years ago, on a trifling wager,
sent a carrier-pigeon from London by the coach to a friend
at Bury St. Edmunds, and along with it a note, desiring that
the pigeon, two days after its arrival there, might be thrown
up precisely when the town clock struck nine in the morning.
This was done; and the pigeon arrived in London, and flew
into the Bull Inn, in Bishopsgate Street, at half an hour past
eleven o'clock, of the same morning, having flown eighty-eight
miles in two hours and a half.

A Pigeon fond of Music.—It is recorded that Mr. John
Lockman, being in the house of Mr. Lee, who lived in

Cheshire, and whose daughter played the harpsichord, observed a pigeon which, whenever the lady played the song of 'Spera si' in Handel's opera of 'Admetus,' would come from the dove-house to the room window where she sat, and listen to it apparently with the most pleasing emotions; and when the song was finished it always returned immediately to the dove-house.

TURTLE DOVES. — This species of dove belongs to the Columbinæ family. They are migratory, and arrive in England usually rather late in the spring, and depart towards the end of August. 'The turtle-dove is about twelve inches long. Its eyes are yellow, encompassed with a crimson circle; top of the head ash-grey, mixed with olive; each side of the neck is marked with a spot of black feathers tipped with white. The back and quill feathers vary in colour,

Turtle Dove.

which, though quiet, harmonize very beautifully. This bird builds in high trees and lays usually two eggs only. During the short time they are here they mate, lay their eggs, and rear their young. On account of their gentle and soothing accents when cooing, and their general deportment, they are considered to be perfect emblems of connubial attachment.'

CHAPTER XI.

FOREST ACROBATS, LITTLE MARAUDERS, AND FLYING ODDITIES.

> Thou lively little nimble thing,
>   Roam freely in thy native wood ;
> Enjoy thy life in quiet there ;
>   *I* would not cage thee if I *could*.
>
> How beats thy heart with fear when human foes
> Thy young ones in their cosy nest disclose.
>
> Though thou art but a link between
>   The mouse that runs, and birds who fly ;
> A bat in nature has its use,
>   A fact 'twere foolish to deny.

MONG the numerous animals confined in small cages and boxes in many of the low and crowded streets of London and other large towns, may be seen one, of a lively temperament, passing its time by turning, with its little feet, a circular part of a wired cage so rapidly that the wonder is the limbs of this creature are not broken by the rotatory motion.

Although absence from its native woods and confinement in so contracted a prison must of necessity lessen the pleasure of its life, it affords by its surprising activity at least some entertainment and instruction to those disposed to learn, but who may never have seen this animal in its state of wild freedom, in which, unrestrained by man, it can follow out its instincts and indulge in its own peculiar habits and tastes. It will at once be seen that we refer to

THE SQUIRREL.—This remarkably light and active animal belongs to the order *Gliris*, and is found in most of the woods and forests of England. Although squirrels are widely distributed over both the Eastern and Western hemispheres, and present many points of similarity in structure, they differ in others. Some are known as flying squirrels, having their limbs invested in a skin or membrane, which they spread out, and by which they are buoyed up and assisted in their movements from one place to another in a manner similar to the birds of the air when flying. Our squirrels have their limbs perfectly free. Being more of a climbing animal than a leaping one, its fore-legs are a little longer than the hind-legs, and have in them a greater concentration of muscular power. This structural arrangement is a great advantage to it in climbing and in laying hold of the boughs of trees when leaping from one to another. As squirrels have to seek their food in the deep shade of thick trees, they

Squirrel.

are favoured with large prominent eyes the more easily to discern it, and to make sure footing in their leaps. Their tails they can bring over the back part of their heads to serve as a kind of shade, and they no doubt act as a balance to the body while springing from tree to tree.

The English squirrel is not only a light, lively, beautiful and active animal, but an exceedingly provident one. In summertime it feeds a great deal upon the young shoots of the pine, of which it is particularly fond. During the autumn it labours assiduously in collecting nuts and various seeds, and storing them in the hollow of some tree as a provision for the winter.

Although these little animals must be as a rule far happier in the wild woods, their native home, than in a state of confinement, however well and kindly treated, they have been known, when tamed, to show considerable attachment to their owners, and to place great confidence in their protection, as well as affording great amusement by their frolics and gambols.

ANECDOTE OF A SQUIRREL.—Captain Brown says: ' A gentleman procured one from a nest, found at Woodhouselee, near Edinburgh, which he reared and rendered extremely docile. It was kept in a box below an aperture, where was suspended a rope, by which the animal ascended and descended. The little creature used to watch very narrowly all its master's movements, and whenever he was preparing to go out, it ran up his legs and entered his pockets, from whence it would peep out at passengers as he walked along the streets, never venturing, however, to go out. But no sooner would he reach the outskirts of the city, than the squirrel leaped on the ground, ran along the road, ascended to the tops of trees and hedges with the quickness of lightning, and nibbled at the leaves and bark ; and if he walked on, it would descend, scamper after him, and again enter his pocket. . . . . This gentleman had a dog, between which and the squirrel a certain enmity existed. Whenever the dog lay asleep, the squirrel showed its teasing disposition by rapidly descending from the box, scampering over the dog's body, and quickly mounting its rope.'

Squirrels have sometimes been robbed of their young by magpies, who have invaded their nests for that purpose.

SQUIRRELS AND NUT-GATHERING.—We copy the following interesting information from the December number of *Nature*, 1876 : 'On the lawn before the window near which I am writing is erected a tripod of three lofty poles, at the summit of which is suspended a basket containing nuts and walnuts. The squirrels, of which there are many in the shrubberies and adjoining plantations, ascend these poles, extract a nut from the basket, and quickly make their way down and across the lawn, in various parts of which they bury their nuts, scratching a hole in the green turf, putting in a nut, filling up the hole, and lastly, with much energy, patting the loose materials with their feet till the filling-up is made firm and solid. This morning for a considerable time only one squirrel was at work, giving me a better opportunity of observing the mode of operation. His journeys were made in all directions, and varied from five feet to nearly two yards, never, so far as I could observe, going twice to the same place, or even nearly so. The squirrels, I am told, forget the spots where they hide the nuts, and in the following spring the lawn, which is very spacious, is dotted with the young plants of nuts and walnuts. As the colours of flowers attracting bees and moths promote fertilization, so the racy flavour of a nut, irresistible to a squirrel, contributes to the distribution of its kind.'

THE MOUSE.—This well-known animal belongs to the same family and order as the rat. There are several species of this animal.

1. There is the wood-mouse, which lives in fields and gardens, and is found in every part of Europe. They commit terrible ravages, and are great enemies to the farmer, the florist, and nurseryman. Baird says : ' They form large magazines of acorns, grain, nuts, etc., for their winter provision ; and the mischief done to the farmer by hogs rooting up the ground is said to be caused by their searching for these subterranean treasures.'

Mouse.

2. There is the harvest-mouse, a tiny creature not more than about two inches long from the nose to the root of the tail, which is two inches more. Their little, cosy, warm nests are usually built amidst standing corn, a little way from the ground, and sometimes in large thistles, which position reminds us of the nests of reed-warblers, which are suspended among reeds in a similar manner. The harvest-mouse is often, with his family, carried to the barn or to the rick, where they multiply to an alarming extent, and commit such devastations that the computed value of the corn is very much less than the farmer expected.

3. There is the house or common mouse, which is found in nearly every place where man is found. They breed rapidly, and cause great destruction of whatever food they may find accessible. They are great invaders of granaries, cheese-mongers' stores, pantries, and larders. They are partial to tallow-chandlers' shops, and whenever practicable visit libraries and book-shops. It is said that a small number of mice are useful in a house that is infested with black beetles, which the mice destroy.

The common mouse is not only a lively interesting animal, but possesses a fair share of intelligence, and has been known to become very familiar with those who have treated it kindly and gently. Drummond says 'that Baron Frederick Trenck, during his long rigid confinement in the fortress of Magdeburg, found a companion in a mouse, which he had rendered so familiar that it would play round him and eat from his mouth. " In this small animal," says he, " I have discovered proofs of intelligence too great to easily gain belief; were I to write them . . . such philosophers as suppose man alone endowed with the power of thought, allowing nothing but what they call instinct to animals, would proclaim me a fabulous writer, and my opinions heterodox to what they suppose sound philosophy." '

This intelligent mouse, which was wont to come at his whistle, jump on his shoulder, and caper on a trencher, was barbarously taken from him, presented to a lady, and put into a cage, where it pined, refused sustenance, and ultimately died, the cause being, no doubt, its separation from the man to whom it had become so ardently attached.

INGENUITY OF A MOUSE TO EXTRICATE HER YOUNG.— We quote the following from the *Animal World :* 'Being troubled with mice, I purchased two traps; one on the spring principle, fastened down by thread, on cutting which, to get at the tempting bit of cheese, poor mousy would be caught and soon put an end to.   But after one or two being caught, it was a most difficult thing to tempt any more of the fraternity to go to the same place.   They appeared to have a thorough knowledge for a long time of its nature and purpose. Noticing this, I purchased a second trap, made of wire, nearly round, with a flat wooden bottom, at the sides of which are placed two or three holes, through which mousy has to push to reach her coveted morsel; but once through, the points of the wires would close upon her to prevent her return.   Poor mousy appears to discover her retreat cut off before touching her prize, because it was generally left untouched, unless there were two or more in at the same time, which appears to lessen their trouble of mind and timidity at being imprisoned.   I have found on one or two occasions that one being alone, although caught in the night, died before morning through fear or some cause which I could not account for; but they were evidently young ones.   Now for the hero-mouse of my paper.   The trap was placed in a kitchen cupboard where it could be seen when the door was open, which happened to be so when this interesting event was witnessed.   A tiny young mouse was seen in the trap, which did all it could to get away, but at every attempt failed.   I was just about to take

pity on the youngster and let it escape, when, lo! an older
one appeared on the scene, evidently the parent. She appeared
to examine the trap all over, and seemed to try to coax her
offspring after her, but to no purpose. At last she left, giving
up her little one, as I thought, for lost; but no, she soon
returned from amongst the rubbish in the cupboard with a
piece of string in her mouth. One end of this she deliberately
pushed between the wires into the cage, and soon made
the prisoner understand what it was to do. Whether the
young one really understood itself, or whether the old one
made it understand by a certain language of their own, I
cannot say; but however, the youngster soon took hold of
the end of the string, and the moment the old one saw she
had a good hold she pulled away with a will, and got her out
almost in a second. The wire at this particular part was a
little more open than in any other part of the cage; whether
this was seen by the old one, or was an accident, is another
problem.

SINGING MICE.—Much has been said about singing mice.
Whether this gift or power arises from a peculiar organization,
or from a functional derangement of their throat or larynx,
may be difficult to tell. In referring to this subject a London
veterinary surgeon says: 'I am sorry to spoil the interest
naturally aroused by such a phenomenon, but I think I have
somewhere seen a similar case, the cause of which was disease
of the air passages, producing constriction, and a consequent
shrill sound with the breathing.'

That instances of mice singing have been known cannot
be doubted, to which we may add the following account taken
from the *Animal World*, in which the writer says:

'A curious case of a singing mouse has lately come under
my own immediate knowledge, and as I think it will be of
interest to you and the numerous readers of your publication,
I hasten to make it known to you.

'About a fortnight ago, the wife of a builder living in Great
College Street, Monte Video Place, Kentish Town, was
startled one evening by seeing a mouse come out of a hole
near the fireplace and run about, singing as it went. The
woman was so alarmed that she fled from the house into that
of her next-door neighbour, who went back with her, saw the
mouse, and informed her it was a most unlucky thing to have

in the house, being a sign of sickness and death. The poor animal was, therefore, poisoned, and thus fell a victim to ignorance and superstition. The woman described the singing as resembling that of a canary.'

THE MOUSE AND THE 'GREENBACKS.'—We give the following curious information of the delinquencies of a mouse on the authority of the *Pall Mall Gazette* newspaper, in which it appeared a short time since. It says :

' There has been a terrible robbery in Ohio, United States. A quantity of " greenbacks " have been stolen by a mouse, who was imprudent enough to eat of them, and died from the effects of the arsenic contained in the green colouring of the money. It seems from the account given of the affair in the Cleveland *Plain Dealer* of the 22nd ult., that Misses Gunson and Middleton are dressmakers and milliners, doing a prosperous business ; indeed, so successful has been their business that they have been enabled in a few months to save 200 dollars to pay the last instalment of the purchase-money of their establishment. This money they stowed away in an old satchel, which they kept under the counter of their shop. The other day, as the time for payment was approaching, one of the ladies looked into the satchel to see if the money was safe. To her horror the satchel was found empty. Suspicion at once attached to a mouse who had lately been seen prowling about the shop. By carefully scrutinizing the floor, the course of the diminutive burglar was tracked by bits of greenbacks to a hole in the wall. The hole was at once searched, and there was found the mouse lying dead in a nest of greenbacks to the amount of 200 dollars. The bills and scrip had been torn to fragments, so that not a cent. could be saved. Much sympathy is felt for the Misses Gunson and Middleton under this misfortune, but none for the mouse, who, although it was neither aware of the mischief it was effecting nor of the danger of greenbacks as an article of diet, is considered to have met with a just retaliation for its inconsiderate conduct.'

BATS, OR FLYING ODDITIES.—These nocturnal animals belong to the class *Mammalia*, Order III. *Canaria*, Sub-order I. *Cheiroptera*, which signifies *wing* and *hand*, or wing-handed. The bat family is widely distributed over the earth. The flying fox-bats are natives of Java, where they are very numerous. They measure nearly five feet in expanse of wing. They are, during the day, comparatively motionless, hanging in such a manner from the trees in which they lodge, that anyone not particularly observant might take them for fruit.

No sooner, however, does the sun go down than they show signs of life and activity by flying direct to villages, orchards, and plantations, and committing considerable mischief to fruit of various kinds. Their flesh is white and delicate, and has a strong smell of musk. It is very probable that at some remote period this species of bat existed in England, as fossil remains of it have been found in our caves.

THE VAMPIRE BAT.—The vampire bat is a native of

Vampire Bat.

South America, and according to accounts given of it by travellers in that country, is a bloodthirsty, troublesome, and dangerous animal. It has been known to attack fowls, quadrupeds, and human beings, and to suck their blood until they have become so weakened as to be scarcely able to stand or to walk, and in some cases until they have died. Waterton says: 'That an unfortunate jackass, attacked by one of these bats, died by inches, and looked very much like misery steeped in vinegar.' They will attack men by biting their toes while asleep, and, without waking them, gorge themselves with their blood. They are, however, capricious, as they will often night after night fall upon the same persons, never attempting to molest others who may be sleeping in the same place. The bats, natives of Great Britain, are known as the large and small horseshoe bats. Like their relations in other countries, they remain, in warm weather, hidden in some quiet place during the day, and come out about dusk, flying hither and thither in a very erratic but rapid manner. They are seen principally during the hot evenings of summer, although they have occasionally appeared in very mild weather in winter-time. Throwing up caps and traps to catch them in their flight constitutes one of the amusements of village boys, although the perpetuation of the effort to secure them is by no means indebted to the success which has followed that effort. It has failed, and no doubt still does fail, in nine cases out of ten.

In remote districts our bats are regarded by superstitious people as omens of evil. Their cry is a peculiar and an un-pleasant one, and is particularly regarded so if heard in a solitary lane, or by the side of some gloomy-looking building or village church. Their noiseless flight, odd appearance, and melancholy tone conjure up in the minds of timid people supernatural existences, and give rise to wild weird imaginings of freebooters, loss of property, and sometimes of life itself. Such is the potent power which superstition has over the human mind in this the nineteenth century of boasted enlightenment and civilization.

STRUCTURE OF BATS.—The peculiarity of the structure of bats renders them a kind of intermediate animal between birds and quadrupeds. They have their forearm converted into a wing by an extension of the membraneous skin which, rising from the sides of the neck, is spread between their fore-feet and their fingers. They can raise themselves and fly like birds, although it is with some difficulty if they have to rise from the earth. We are told that the inflation in birds is made by direct communication with the lungs. In bats a small opening is at the bottom of the cheek-pouches, and is furnished with apparatus by means of which air is prevented from escaping without the will of the bat. When the bat desires to inflate the body it closes the mouth, and forces the air into the empty space between the skin and the flesh. It then looks like a ball of fur, to which its head and limbs seem artificially attached. The Jews were forbidden to eat bats. 'The Greeks took them as the models of their disgusting harpies or monsters, with the face of a virgin, the body of a vulture, and the wings of a bat. In the Middle Ages they were made the companions of sorcerers and evil-disposed men, and when they wished to represent Satan they placed a pair of bats' wings on his shoulders.

Uses of Bats.—In proportion to the minuteness of those insects whose numbers are so great, and whose voracity is so destructive to corn, fruit, flowers, and vegetation, so man is at a loss in devising means by which to destroy or keep these insects in check. Every agent therefore provided by Nature to live upon these small creatures, so as to lessen the depredations they would otherwise commit, should be regarded as the real friends of man, amongst whom bats may be included. When bats are on the wing they are in search of insects, on which, in England, they principally feed. We may, therefore, consider them as useful co-workers with insectivorous birds and moles in preserving our crops.

In some countries the natives make soup of them, which is said to be not only palatable, but also nutritious. Bats have been known to remain without food considerably longer than the period during which any human being has ever been known to fast. In the autumn bats usually become very fat inside their bodies. It is by this fat that, during hibernation, they are kept alive. When the winter is short they emerge from their places of seclusion in tolerably good condition; but if the spring is backward, cold, and wet, they become very weak, and some of them even die.

CHAPTER XII.

FEEBLE FOLK, FISHERS, AND POACHERS.

'She flies, she leaps, and bounces to deceive,
  Till fainting, breathless, spent, at last she drops,
  On some fresh verdant turf, or thymy bank,
  Once the gay scene of her nocturnal sports.'

T is interesting to note that animals differ in their habits, tempers, and dispositions according to the nature of the food on which they subsist. Carnivorous animals who feed only on flesh are more fierce, dangerous, and destructive than all others, especially when overtaken by hunger. To meet a lion or a pack of wolves at such a time would imperil the life of the bravest man.

Farmyard fowls, pigs, and other omnivorous animals, who will eat flesh, worms, seeds, and vegetables, or almost anything that is eatable, are less savage than lions and tigers. Grain-eating birds are not so bold or fierce as eagles and vultures, whose food is the flesh of other animals, and who have been known to attack and carry them alive to their eyries to be devoured at their leisure.

Herbivorous animals, such as horses, oxen, sheep, and we may add hares and rabbits, and others belonging to the same family, are comparatively harmless; at least they are not aggressive on other forms of life for food, and only when necessary place themselves on the defensive against their tormenters, or those animals that might attempt to destroy them.

'Feeble folk,' included in the group of animals we have selected as subjects of this chapter, will first have brief attention.

HARES AND RABBITS.—These animals are so well known as to require but a short description. They are rodent animals belonging to the class Mammalia, have sharp cutting teeth, large eyes, and long ears, varying, however, according to the species.

Referring to hares, Baird says: 'They are gentle, timid animals, easily frightened at the least noise, and are possessed of remarkably quick hearing. Their mode of progression is by leaping, and when alarmed their flight is very rapid.' They are very prolific, and would become very mischievous if they were not kept in check by numerous enemies, such as weasels, foxes, and similar canaria. The hare does not burrow, but simply hides under a bush or in a furrow,

Hare.

and such place is called its *form.* The young are born with their eyes open. Hares constitute an important article of food in this country, although both the Mahometan and Jewish religions prohibit their followers from eating them.

Coursing or hunting such a small, timid, and defenceless animal as the hare may afford pastime and sport to man, but it certainly does not add to his dignity or show either courage or heroism. Probably lion or tiger hunting might do so.

COWPER'S HARES.—Wild as hares are in their natural condition, they are capable of being sufficiently tamed so as to become interesting and attached companions of those who treat them kindly. The affection of Cowper the poet for his three tame hares, Bess, Tiny, and Puss, is well known. Puss became more attached to his master than to his own species. On fine days he was always anxious for his master to go into the garden, and would give signs of his wish by pulling the poet by the skirts of his coat, drumming on his knee until he had fairly got him out. Of Tiny, Cowper gives the following account: 'He was very entertaining in his way; even his surliness was matter of mirth, and, in his play, he

preserved such an air of gravity, and performed his feats with
such a solemnity of manner, that in him, too, I had an agreeable companion.' Bess was the comedian and acrobat of the
company : 'A hare of great humour and drollery.' Puss was
tamed by kindness. Tiny refused to be tamed at all. But
Bess 'had a courage and confidence that made him tame
from the beginning.' The poet used to treat his pets to a
'carpet-dance' in the lonely evenings, on which occasions,
'Bess being remarkabably strong and fearless, was always
superior to the rest, and proved himself the Vestris of the
party.' Bess was cut off in the prime of life. Tiny lived to
be nine years old, and Puss survived him two years.

REMARKABLE STORY OF A TAMED HARE.—'We are informed,' says the author of 'Anecdotes of the Animal Kingdom,' 'by Borlase, in his "Natural History of Cornwall," that
he had a hare so completely tamed as to feed from the hand ;
it always lay under a chair in the ordinary sitting-room, and
was as much domesticated as a cat. It was permitted to take
exercise and food in the garden, but always returned to the
house to repose. Its usual companions were a greyhound
and a spaniel, with whom it spent its evenings. The whole
three seemed much attached, and frequently sported together,
and at night they were to be seen stretched together on the
hearth. What is remarkable, both the greyhound and spaniel
were often employed in sporting, and used secretly to go in
pursuit of hares by themselves, yet they never offered the
least violence to their timid friend at home.'

THE HARE THAT LOVED MUSIC.—The same author says :
'There is an anecdote related of five choristers, who, while
singing an anthem by the banks of the Mersey, in Cheshire,
attracted the notice of a hare ; when they ceased she made
off, but on their again commencing she returned quickly, and
stood about twenty yards distant in the open field. When
they finished she again bent her way to a neighbouring wood.'

HAZARDOUS VENTURE BY A HARE.—Fouilloux says he
saw a hare start from its form at the sound of the hunter's
horn, run towards a pool of water at a considerable distance,
plunge in, and swim to some rushes in the middle, where it
lay down and concealed itself from the pursuit of the dogs.

A BATTLE BETWEEN TWO HARES. — Mr. Waterton, in his
'Essays on Natural History,' says : 'One Easter Sunday, in

the afternoon, as I was proceeding with my brother-in-law, Mr. Carr, to look at a wild-duck's nest in an adjacent wood, we saw two hares fighting with inconceivable fury, on the open ground about 150 yards distant from us. They stood on their hinder legs like two bull-dogs, resolutely bent on destruction. Having watched them for about a quarter of an hour, we then entered the wood; I observing to Mr. Carr that we should find them engaged on our return. We stayed in the wood some ten minutes, and on leaving it saw the hares still in desperate battle. They had moved along the hillside, and the grass was strongly marked with their down for a space of twenty yards. At last one of the sylvan warriors fell on its side, and never got upon its legs again. Its antagonist then retreated for a yard or so, stood still for a minute, as if in contemplation, and then rushed vengefully on the fallen foe. This retreat and advance was performed many times; the conqueror striking its prostrate adversary with its fore-feet, and clearing off great portions of down with them. In the meantime the vanquished hare rolled over and over again, but could not recover the use of its legs, although it made several attempts to do so. Its movements put you in mind of a drunken man trying to get up from the ground after a hard night in the alehouse. It now lay still on the ground, effectually subdued, while the other continued its attacks upon it with the fury of a little demon. Seeing that the fight was over, we approached the scene of action; the conqueror hare retiring as we drew near. I took up the fallen combatant just as it was breathing its last. Both its sides had been completely bared of fur, and large patches of down had been torn from its back and belly. It was a well-conditioned buck-hare, weighing, I should suppose, from seven to eight pounds.'

Although the conduct of the two hares referred to in the above anecdote appears to be altogether inconsistent with the natural gentleness and timidity of these creatures, it is nevertheless true that they, and other tribes of animals, as well as human beings, are not only capable of feeling the passion of love, but even of jealousy one towards another.

Thompson says: 'The males of deer and cattle are extremely tenacious of their rights, and engage in instant battle with the trespasser.'

It may be reasonably assumed that both the hare combatants spoken of were males, and that the cause of the quarrel between them was an attempted undue interference by one of them with the connubial rights of the other. Or it may have arisen from the one having invaded what the other considered to be its own exclusive domain, and that the latter animal, to punish the offender, acted on the same principle as that adopted by the scavenger dogs of Constantinople, who summarily castigate a strange dog who may attempt to pick up a living in the same street or locality. Whatever the real cause may have been, we have in this battle of the two hares a very marvellous phase of animal life.

CLEVER HARES.—In addition to the proofs furnished in the interesting account given of Cowper's hares that these animals are capable of education, we give the following information. It appears that at one time a hare was exhibited in England which not only danced to measure, but, to the astonishment of beholders, played with its fore-feet upon a tabaret, and observed a correct number of strokes. Afterwards the hare fought with a dog (no doubt instructed for the purpose), when it bit with its teeth and beat forcibly with its feet. Another hare, many years ago, was taught to beat a drum with its fore-feet while a person carried it round the stage.

AN AMUSING COLLOQUY.—If village boys are not as a rule as quick and confident in their answers to interrogations as town boys may be, there are many exceptions to this rule, as some have been known to show much shrewdness, and quaintness too, in the replies they have given to certain questions, of which we have a proof in the following amusing story, contained in the 'Parlour Menagerie': 'A sportsman coursing having lost his hare, thus hastily accosted a shepherd's boy: "Boy, did you see a hare run by here?" "A hare, sir?" "Yes, fool!" "What, a hare, sir?" "Yes!" "What, a thing that runs fast, with long ears?" "Yes." "That goes lopperty, lopperty lop?" "Yes, yes, my good fellow!" "What, very long ears?" "Yes, dolt!" "Ah, then," said the boy, "I didn't see it."'

RABBITS.—Rabbits differ in their habits from hares, inasmuch as they are essentially burrowing animals, and live in companies of hundreds and sometimes thousands in the same wood or warren, while hares are comparatively isolated, and content themselves with the surface of the earth. Rabbit-burrows are often very long and irregular. Into these they run in times of danger, but even there they are not always safe, as they have been followed into them by stoats, weasels, and ferrets, and there destroyed. Foxes and hedgehogs often indulge their carnivorous appetites with a rabbit, and birds of prey have been known to pounce upon them and to carry them off. They are truly a 'feeble folk.' They have no means of defence to ward off the rapacious animals just mentioned.

Rabbits are not without ingenuity, as they have been known to watch a terrier dog into a burrow, then to fill up the entrance with earth, so that the invader could not escape, but has perished miserably underground. They breed seven times a year, so that a single pair in four years may become the progenitors of 1,274,840 descendants. A military force from Rome had to suppress the numbers of these animals, which overran the islands of Majorca and Minorca.

Rabbit.

The following information, taken from the *Evening Standard* newspaper of the 9th October, 1882, may be interesting to our readers, especially to those of them who take an interest in the animals to which reference is made :

'It has generally been supposed that the rabbits brought over in such enormous quantities to this country from Ostend are wild rabbits, or, rather, rabbits bred in vast numbers in great warrens. Attempts have been made here, with more or less success, to breed rabbits on a large scale in the same manner; but the success which has attended these efforts has certainly not been sufficiently marked to produce a supply which can compete, in either number or cheapness, with the rabbits shipped from Ostend. The *Provisioner* states that, with a view to find out how so cheap and plentiful a supply was raised, it has investigated the rabbit question upon the

spot, and finds that the popular idea is a wholly mistaken one.
The rabbits are neither wild nor raised in warrens, but are of
the ordinary domestic description with which all of us are
familiar, and are bred by the farmers and labourers in the
agricultural districts of Belgium and France.   They can
hardly be said to form an item in the various occupations
which are engaged in by farmers for profit, since, although
they are bred with the ultimate intention of selling them, they
are nowhere cultivated in such numbers or with that careful
attention which characterize other pursuits forming a staple
industry.   It is even a common thing for them to be bred in
twos and threes by young children.   The rabbits are collected
periodically by higglers, who go round from place to place in
vehicles, and buy such as are then in a condition for killing.
Rabbits, in fact, are raised on a small scale in Belgium, as are
eggs in France, as a sort of by-product, and it is, under these
circumstances, astonishing indeed that such an enormous
number of these creatures can be sent to England at so cheap
a price.   The magnitude of the trade may be inferred from
the fact that in many days of what is known as "the season,"
extending from October to March, as many as two thousand
boxes are shipped from Ostend alone.   As each box contains
on an average ten dozen rabbits, this gives a total of nearly
a quarter of a million at one shipment.   Surely what is done
in Belgium might be done in England.   If all our cottagers
could but be persuaded to keep a few hens and a few rabbits,
they might materially add to their earnings, and the country
would save the very large sum which she yearly pays to her
neighbours for provisions which might be profitably raised at
home.'

Rabbits have their tempers, which they have a curious way
sometimes of showing.   In a domesticated state they are easily
offended, and often give way to sudden starts of rage by
thumping loudly on the ground with their hind-feet.

'It may be observed,' says Brown, 'that the rabbit, like the
other tribes of the hare, though so harmless to other animals,
frequently fights with its own kind.   Two males, confined in
the same place with the other rabbits, will be sure to fight,
and the stronger will bite and persecute the weaker incessantly.
The effect may often be seen in wounds, extending over the
back of the animal.   The rabbit will also, when offended,

bite the hand or leg of the person nearest it with its sharp front teeth. It may be noticed, too, that the male rabbit has a great propensity for destroying the young; to prevent which, the female carefully covers up the nest each time she goes out to feed, and when domesticated, seeks a place of concealment from the male.'

From 'All about Country Life' we select the following information, which may be useful to

BOYS WHO KEEP TAME RABBITS.—'Tame rabbits are often kept for profit or amusement in hutches or houses built on purpose for them. There are many sorts of common tame rabbits, of great variety in size, colour, and appearance. Some are entirely black; others white with red eyes; there is also the silver-grey or ash-coloured rabbit, and many sorts of variegated colours. Then there are the fancy rabbits, of which the most celebrated are the "dew-lop," the "oar-lop," the "flat-lop," the "half-lop," and the French rabbit. Fancy rabbits are usually of large size, and a fine one will weigh twelve pounds. They were originally brought from the East, and require more warmth than common English domestic rabbits.

'Perfect cleanliness and suitable food regularly given form the essential conditions for success in rabbit-keeping. Half of the diseases of tame rabbits arise from not keeping their hutches cleaned out sufficiently often, and a large proportion of the other half from injudicious feeding. If fed exclusively on green, succulent food, they are apt to get pot-bellied. No vegetables should be given them in a wet state. The mistake must not be made, however, of feeding with too much corn or other dry food. Too many oats will kill rabbits faster than too large a quantity of grains. The true secret of good keeping lies in a varied diet. Carrots and hay are extremely useful in winter, and in summer many wayside and garden weeds are peculiarly wholesome. Among them the sow-thistle, the dandelion, the plantain, and groundsel recommend themselves, while parsley, endive, celery-tops, lettuce and cabbage-leaves may be gathered from garden plants, and the various cultivated grasses from the field. The stalks and leaves of chicory are said to be excellent for rabbits.'

To the above directions we would offer a suggestion which, if adopted, would no doubt help to ameliorate the monotony of what is after all a life of imprisonment to these animals.

We refer to the necessity of giving them more room in their hutches or houses than is usually allotted to them. The young of tame rabbits, as well as those of other animals, are fond of play, which they would be able to indulge in in proportion as they have space to exercise themselves, and which would really be promotive of their pleasure, health, and growth.

OTTERS.—These animals are found in almost every part of Europe, and at one time were common in England, especially in the northern part of Yorkshire; but their numbers have

Otter.

been so decimated by otter hunts, that they are now but rarely met with.

The otter belongs to .the class Mammalia, family Felidæ. It is an aquatic and amphibious animal, and is found principally on or near rivers and streams where fish abound, and on which it mainly subsists. It is quite an epicure, as it has been known to eat the middle of the back of a fish, and to reject the head and tail parts, which have been picked up by hungry peasants, taken home, washed, and eaten. The body of the otter is long, its legs are short, and its feet are webbed, tail flat, and physiognomy very remarkable. Its structure is well adapted for its aquatic life, but it has an awkward gait on land. Its toes are united with membranes in the same way as the feet of ducks and geese. When its feet go forward the toes are closed, but when thrown backward they spread out in order to get well hold of the water, and so propel the animal onward. As fish always when feeding direct their heads up the water, otters go against the stream, so that they may come on the fish unawares. In returning to their burrows with the fish, they can the more easily convey it down the stream than against it.

The home or nest of the otter is a bed of rushes or similar material under hollow banks by the river's edge, where they produce their young. The English otter usually hunts for its prey by night, and if not disposed to take to the water, feeds upon snails and worms, which come out at the same time. This animal may be considered as the link between the seal and the quadruped. Its skin is very valuable, and is much used by furriers. In some continental countries its flesh may be eaten in the time of Lent.

SEA OTTERS.—The sea otter is much esteemed on account of its fur. The American otter is about five feet long, with fur of a glossy brown. 'About 8,000 skins of this species are imported into England every year. The South American otters live in troops, rise to the surface of the water, and bark like dogs in a menacing and snapping manner.'

Although otters are as a rule very untractable, they have been tamed, and rendered, while living, useful to man. A writer in Heber's Journal says : ' We passed ten beautiful otters, tethered with straw collars and strings to bamboo stakes, on the banks of the Matter Colly; some swimming, some half out of the water, some basking in the sun, and making noises as if in play. Fishermen keep them to drive shoals of fish into their nets. Sometimes they would bring the largest fish with their teeth the same as dogs.'

Otters show strong affection for their young. Lieutenant Wood, in his journey to discover the source of the Oxus, disturbed a colony of otters and secured two young ones, which were put into a sack. Their cries brought the old otters round the boat all night. They followed the boat ten miles, and even attempted to get in. So strong was their affection, that they were altogether fearless of danger.

It is said that the otter and seal have a faculty in the eye, by means of which they are able to elongate or shorten the axis of the organ at pleasure, and by that means to see equally well in two media of very different density; namely, water and air.

THE WEASEL.—In the habits and dispositions of many animals, we have true types of the moral and even general character of some human beings. The fidelity and attachment of the dog remind us of those men whose sense of honour and duty is an incentive to an open, straightforward, and honest course of life; while the habits of the sly, stealthy, and voracious weasel kind remind us of men whose love of self, want of good principle, and evil dispositions, would sacrifice the rights of others, and rob them of any benefit or advantage they could, simply to accomplish their own objects, and to promote their individual pleasure and interest.

But what is the true character of the weasel? Voracious and cunning though he may be, he is not an 'unmitigated villain,' but has some good traits of character, and is of some use, to recommend him to our notice.

The weasel tribe belongs to the order *Feræ*, and comprises several kinds. There is the *striated weasel* of America, noted for the very disagreeable odour it has the power of emitting when irritated. The *honey weasel*, so called because of its fondness for honey, is found in Southern Africa. This animal looks out for the bee returning to its hive, and should he find one within his reach he at once attacks it. It has no fear of the sting of the bee, because its skin is too tough for the sting to perforate it. The *ichneumon* belongs to the same tribe, and is also found at the Cape of Good Hope, in Barbary, and Egypt.

Ferret.

FERRET AND POLE-CAT.—The *ferret* is also a native of Africa, and is much used in England by gamekeepers and ratcatchers for catching rabbits and rats. It emits a very offensive smell, particularly when angry. It has red eyes. The polecat, or foumart, is not uncommon in our own country. Its skin, when taken in winter, is esteemed very valuable.

The stoat, or ermine, is also an inhabitant of England, and is a little longer than the common rat. In summer its coat is brown ; in winter, white. It is then, on that account, usually taken. Its fur is so highly esteemed that it has been from remote times and to the present worn by royalty.

This animal is particularly fierce and determined in its efforts to secure its prey. They pounce upon rabbits with amazing quickness, suck their blood, and devour their bodies at leisure. We have read of a gentleman releasing a rabbit from the clutches of a stoat, which unwillingly gave up its prey, as it followed the gentleman, and sprang at the rabbit while he was carrying it by the legs.

The common weasel is about six or seven inches long, and is found in places near the habitations of man. It has a smallish head, bright quick eyes, short legs, and a long neck. It is exceedingly agile, and can twist its body about as lithely as one could a thin strap of leather. Its leaping powers are considerable, as it can bound several feet when alarmed by an intruder ; and it can spring from branch to branch, and from tree to tree, with the nimbleness of a squirrel.

This little animal is thoroughly carnivorous, and is as much a beast of prey as the lion or tiger. It lives upon moles, rats, field-mice, and birds. When pressed by hunger it will attack animals twice its own size, which it seldom fails to kill. It is useful to the farmers in killing mice and rats which infest their cornricks and granaries. Like the stoat, the weasel will spring upon a hare or rabbit when asleep, fasten upon their neck, and drink their blood while they are running until they drop down dead from sheer exhaustion.

THE WEASEL AND FARMER.—We once saw a weasel stealthily creep up to a young farmer who was soundly sleeping in an old shed in one of the fields. The weather being very hot, the young man was lying with his neck and throat quite uncovered. The nose of the weasel was close to the side of the neck, and he was no doubt just going to commence his practice of phlebotomy, or blood-letting, when we frightened him off, and no doubt deprived him of an anticipated sanguinary meal.

WEASEL WORSHIP.—We learn from Sonnini that the Turks

as well as the Greeks pay great respect to the weasel. It was formerly worshipped in the Thebais. 'The Greek women carry their attention so far as not to disturb it; and they even treat it with a politeness truly whimsical. "*Welcome*," say they, when they perceive a weasel in their house; "come in, my pretty wench; no harm shall happen to you here; you are quite at home; pray make free," etc. They affirm that, sensible of these civilities, the weasel does no mischief; whereas everything would be devoured, add they, if they did not behave to this animal in a courteous manner.'

WEASEL SAGACITY.—'The workmen in the engine-house of the New Haven Railroad were greatly amused by the movements of a weasel that had killed a rat, nearly as large as himself, in one of the engine-pits. The side of the pit being perpendicular, and the rat too heavy for the weasel to carry up in his teeth, the question arose how he should get him out. It looked like a difficult task, but the weasel was equal to the emergency. After several unsuccessful attempts to shoulder the rat and climb up the side, he laid him down and went about to the different corners of the pit on a tour of inspection. Finally selecting one in which sufficient dirt had accumulated to make an elevation of several inches, he went back, dragged the rat to the corner, and stood him upon his hind legs. He then clambered out of the pit, and, going to the corner where he had left the rat, let himself down by his hind-feet from above, clasped the rat round the neck with his fore-paws, pulled him up, and trotted off with him to his hole. The weasel is one which made his appearance at the shops some time ago, and which, by being unmolested, has become quite tame.'

CHAPTER XIII.

BRISTLY PACHYDERMS, WILD AND TAME.

The pig is ugly, some suppose,
Because of his long ears and nose;
They like him, too, or I'm mistaken,
For such folks eat both pork and bacon.

OING back in our own country's history long before our mountains were tunnelled, our valleys invaded, our fields and woods cut through, or our rivers crossed to make way for railway trains; when the towns and villages of England were neither so numerous nor thickly populated as they are now; when the silence of rural districts was broken only by singing and chirping birds, or the peculiar noises of the wild quadrupeds which roamed at pleasure; when there were no telegraphic appliances, and no penny postage for the transmission of newspapers, or of letter correspondence, such as we are favoured with at the present time; when travelling facilities were confined to stage coaches going at the rate of thirty or forty miles a day,—we find it was then that in our numerous, thick, and extensive forests wild swine found homes in which to live, to breed, and rear their young, and the hunting of which afforded popular sport and pastime to many of our countrymen who lived during that remote period.

Whether that custom was a commendable one or not, or even a necessity to keep down the too rapid increase of wild boars, we will not presume to say. One thing is certain, however, that hunting these animals exposed the hunters to much more danger than does the hunting of the smaller and much less formidable animals, the fox and the hare, both of which seem to be, at the present time, the substitutes, in this respect, for the now extinct wild boars of our native land.

THE WILD BOAR.—Before referring in detail to one of our most useful animals, the domestic pig, we may notice that its origin is no doubt identified with the wild animals to which reference has already been made.   Of this we think there can be no question.   In fact, it is quite as easy, if-not more so, to prove that this is the case, than to trace the parent stock of the domestic dog or cat.

Speaking of wild boars, Baird says : ' The wild boar no longer exists in Britain, though in former times it was common enough.   It is found in many parts of Europe, however, as well as in India, harbouring in the most solitary places in re-tired forests.   His lair is generally in some wild and remote spot, not far from water, and commanding by some devious path access to the open country.   As a beast of chase, the wild boar is still held in high repute in some parts of Europe, and in India hog-hunting forms one of the most exciting wild sports that is known.'

The following interesting description of the wild boar is given in the *Animal World:* ' The wild boar is a fierce, hardy animal, which comes to maturity in five or six years; but it sometimes lives twenty-five years, and increases in size, passive strength, and hardihood of character during that time. The young of this wild race are not only attached to their mother, but to one another; and this attachment does not cease when the mother's protection is no longer needed, but is communicated to every fresh litter, till the produce of one mother forms a little colony, the members of which appear capable of recognising one another even after they have been separated for some time.   It is possible that most herds of social animals are originally formed on this principle.   Herds of wild swine are always under the leadership of a male. They assemble and sally forth from the forests, and do no small damage to cultivated fields, both by rooting up and by trampling down.   There is an allusion to this in the 80th Psalm : "The boar out of the wood doth waste it, and the wild beast of the field doth devour it." This is finely true to nature, even in the contrast of the mischief done by the two.'

FOOD OF WILD HOGS.—'The food of the wild hog consists chiefly of roots and vegetables, and, that he may the more easily obtain these, his nose is furnished with a powerful muscular and cartilaginous apparatus. Worms, acorns, beechmast, chestnuts, etc., are greedily sought after and devoured. The position of his tusks defends the eyes in rushing through the underwood. The habits of the wild boar are nocturnal, and, unless aroused by hunters, he will not quit by day the clean dry lair he always forms for himself in the forest; nor will he, if unmolested, attack man, though when provoked his defence is most resolute. A horse that has been once wounded by a boar can never again be induced to approach him. Yet when reduced to captivity, the wild boar becomes comparatively gentle and manageable.'

Hogs are pachydermatous animals, belonging to the order Ungulata. They have four toes to each foot; two in the front, shod with stout hoofs, and two behind, or lateral toes, which scarcely touch the ground when the animal is walking.

As the domestic pig is partial to roots and vegetables, like the wild boar, he is supplied with a prolonged cartilaginous nose, strengthened at the end by two bones, which enable him to turn up the soil with comparative ease; he has also a very strong muscular neck, which is of essential service to him in his rooting operations.

VARIETIES OF THE DOMESTIC PIG.—The species are not very numerous, but they have been arranged in four or five genera. 'It has been asserted,' says the *Animal World*, ' that there exist only three actual varieties of the domesticated hog: the " Berkshire," " Chinese," and " Highland ;" but we are really indebted for our numerous varieties not only to these three well-known races, but also to the " African," " Spanish," " Portuguese," and " Italian ;" chiefly, however, to the wild boar of the European forests.'

From the above periodical we gather the following information respecting the many kinds of hogs well known in England, which are distinguished from each other by a difference in size, the quality and length of their hair or bristles, and by peculiarities of colours, or mixtures of colours. That journal says : ' The *Berkshire* hog is of large size, of reddish-brown colour, with black spots or patches. The *Irish* " *Greyhound* pigs " were a tall, long-legged, bony race, their throats furnished

with pendulous wattles. These gaunt creatures are now seldom seen out of Galway. The *Suffolk* were formerly a small, delicate breed; now the breed is notable for great length and small limbs, etc. The *Cheshire* breed is white, black and white (in large patches), or grey in colour. It is chiefly remarkable for gigantic size. *Hampshire* pigs are usually dark spotted, sometimes black or white. They have been known to weigh forty stones each. The *Yorkshire* breed were formerly large, coarse, heavy animals; now their value consists in symmetry, fatness, and early maturity. . . . The peculiarities of the improved race are small bone, head, and face, deep capacious chest and throat, and neck rising roundly behind the ears, shoulders thick and round, and hams square down to the elbow. *Bedford* swine become enormously fat, grow rapidly, and fatten fast in proportion to the food given. The original *Shropshire* and *Wiltshire* breeds were inferior, but crossings with *Berkshire*, *Chinese*, and *Neapolitan* breeds have resulted in a race compact in form and excellent in quality. *Herefordshire* and *Gloucestershire* are good store pigs ; the latter are white, with wattles hanging from each jaw ; they are hardy, and more profitable for pork than for bacon. The *Northamptonshire* pig is of light colour and handsome shape; he is a profitable porker and good store pig. The best *Norfolk* swine are white; they are short, compact, excellent porkers. A spotted variety of larger size is inferior in point of delicacy. The *Lincolnshire* breed is white, with fine skins, sparingly covered with slender bristles, ears erect and pointed, body long, straight, and round. The recent improvement in the *Essex* breed is due to a cross with the Neapolitan. It is almost bare of hair, and deep jet black in colour ; a very quick feeder, with small bones and great fattening power. The Essex half blacks are smaller, and make excellent meat. *Sussex* hogs are generally black and white in large patches ; they seldom exceed twenty stone in weight, but the flesh is very good. The original *Old English* breed was heavy and bony, but is now, except perhaps in Cornwall, extinct.

‘A very peculiar breed of swine exists in the Orkney Islands, Hebrides, and Shetland Islands. The hog has been described as a “little, ugly, brindled monster, an epitome of the wild boar, yet hardly larger than an English terrier.” These little swine are uncared for by their owners, have no shelter but such as they can find under bank or bush, and depend for food on what their own ingenuity enables them to secure ; but ranging about undisturbed, gain sufficient sustenance from roots, worms, potatoes, plovers' and curlews' eggs, an occasional lamb, etc., to keep themselves in good condition. If put up to feed they fatten on inexpensive food with great rapidity, and increase astonishingly in actual size. These pigs are generally converted into pork, and form no trifling article of commerce between the islanders and coasters. The average weight of these animals is from sixty to seventy pounds. The ropes used to suspend over the cliffs the adventurers who make a livelihood by taking sea-birds' eggs are made from the bristles of these swine ; they are found less liable to cut from friction at the edges of the rocks than ropes made of hemp.'

GIANT AND DWARF PIGS.—Formerly it was a common thing to see at our fairs a caravan with a painting at the outside representing a fat pig weighing fifty, sixty, and even seventy stone. One was exhibited, some years since, in a northern town, and stated by the proprietor to weigh eighty stone. Mr. Knell, near Maidstone, once possessed a dwarf-pig seven inches long from snout to tail, five inches and a half round the body, three inches and three quarters high, and weighing only fourteen ounces. It was one of a litter, all the rest of which were the ordinary size ; it was in good health, squeaked loud, and ran fast.

PIGS NOT EPICURES.—Nothing in the shape of food comes amiss to the domestic hog. He has an appetite for the most impure things, and has been known to devour his own progeny. He has a stomach of enormous size, which no doubt accounts for his great voracity and constant craving for food. His rough, wiry hair, thick fat, and hard skin render him less susceptible to pain from a blow than many other animals are. We are told that mice have been known to lodge upon a hog's back, to eat his skin and his fat, without his showing any signs of sensibility.

Although we often hear it said of men and women who are not cleanly in their habits that they are as ' dirty as pigs,' it must not be supposed that swine do not enjoy a clean, comfortable habitation as well as other animals. They not only do so, but thrive much better when kept dry and clean than when neglected and deprived of the comforts referred to.

We well remember a gentleman purchasing a small pig weighing about five stone of a neighbour; putting it into a properly constructed sty with a sloping floor, giving it plenty of clean straw, and feeding it well during twelve weeks. It was then killed, and the carcase weighed seventeen stone, having gained two pounds of flesh per day during the time he had it.

HOGS AND EGYPTIANS.—We are informed by Smellic that 'the hog has been noticed from the earliest antiquity; one nation has ever considered it as an object of the greatest detestation, while others have ranked its flesh among their principal delicacies. It is well known that the Mosaic law forbids the use of swine's flesh, and such is the abhorrence in which these animals are held by the Jews, that profane writers have rallied them on the subject. The Egyptians were allowed to eat pork but once a year; this was on a festive day, dedicated to the moon, when great numbers were sacrificed to that satellite. If at any other time an Egyptian was unfortunate enough to touch a hog, however slightly, it was necessary for him to plunge into the *Nile* with all his clothes about him, in order to purify himself from the abomination. Those who kept hogs were rejected by society; they were not allowed to enter the temples, nor even to form any alliance with other families. This aversion to hogs has been transmitted to the modern Egyptians. In the island of Crete pigs were held sacred, and in Rome the swine-feeder spared no pains to make them delicious meat.'

The religion of Mahomet forbids the use of swine's flesh because it was considered unclean, and there can be no doubt that it was on this account the Chinese objected to Mahometanism, they being extremely fond of pork and great breeders of hogs.

THEIR USES: BRISTLES AND SKINS OF HOGS.—Domestic hogs are exceedingly useful animals. The females usually produce two litters a year, each numbering from seven to fourteen young ones; this they would, if permitted, continue to do during fourteen or fifteen years.

The males have been compared to the miser, who is useless and rapacious during his life, but whose death becomes a public service. To a nation like Britain the flesh of these animals is a great boon, because it takes salt more easily than any other, is more effectually preserved, can be kept almost any length of time, and eaten in any latitude of the world.   To the crews of our navy and merchant ships bacon is of the greatest importance.   Hog-skins, when properly dressed, are useful for making slippers and the seats of saddles, and are also employed by various artificers.   The bristles are extensively used by brushmakers and shoemakers, as well as by saddlers.   A great trade in bristles is carried on between Russia and this country.

Pigs may be considered to be essentially the poor man's friend, particularly in Ireland, where they are fed in great numbers with a view to double duty—namely, for home consumption and for sale, to *pay the rent.*

INTELLIGENCE OF PIGS.—Hogs are not altogether destitute of sagacity, nor is their sense of smell particularly defective.   We have read of one having been taught to perform the duty of a pointer in hunting, and to do it almost as steadily and effectively as a well-bred dog.   This animal belonged to Sir Henry Mildmay, and was trained by his gamekeeper.

ANECDOTE OF A PIG AND A DOG.—If the following story (which we quote from that excellent work 'Our Dumb Animals') be authentic it would appear that even hogs possess some degree of reasoning powers : ' A pig and dog on board a ship on the voyage from India to London were very good friends; they ate out of the same plate, walked about the decks together, and would lie down side by side under the bulwarks in the sun.   The only thing they quarrelled about was their lodging.   Toby, the dog, had a very nice kennel; the pig had nothing of the sort.   Now piggy did not see why Toby should be better housed of a night than he.   So every night there was a struggle to see who should get into the kennel first.   If the dog got in he showed his teeth, and the other had to look out for other lodgings; if the pig gained possession the dog could not turn him out, but waited for revenge next time.   One evening it was very boisterous, the sea was running high, and it was raining very hard.   The pig was slipping and tumbling about the decks; at length it was so unpleasant that he thought the best thing he could do was to go and secure his berth for the night, though it yet wanted a good time to dusk.   But when he came to the kennel there was Toby safely housed; he had had the same idea as to the state of the weather as the pig. "Umph! umph!" grunted the pig, as he looked up to the black sky; but Toby did not offer to move. At last the pig seemed to give it up, and took a turn as if to see where he might find a warm corner for the night. Presently he went to that part of the vessel where the tin plate was lying that they ate their victuals off. He took the plate in his mouth, and carried it to a part of the deck where the dog could see it, but some distance from the kennel; then, turning his tail toward the dog, he began to make a noise as if he was eating out of the plate. "What!" thinks Toby; "has the pig got some potatoes there?" and he pricked up his ears, and looked hard toward the plate. "Champ! champ!" goes the pig, and down goes his mouth to the plate again. Toby could stand this no longer—victuals! and he not there! Out he ran, and, thrusting the pig on one side, pushed his cold nose into the empty plate. The pig turned tail in a twinkling, and before Toby knew whether there was any meat on the plate or not, piggy was snug in the kennel. Of which we may add, he no doubt kept possession all the night.'

Although the word 'gourmand' is appropriately applied to the hog because of his voracious appetite, we hope the foregoing remarks conclusively prove that he is a *useful 'gourmand.'* Pigs are by no means devoid of intelligence, and they may, like many other animals, become not only domesticated, but, by kindness and humanity, familiar companions of man.

A NOTED PIG IN CHELSEA.—Two summers ago we frequently saw, in the King's Road, Chelsea, a small pig with a blue ribbon frill round its neck and a half kind of jacket on its body, and accompanied by a young man, its master. This pig would thread its way among the foot-passengers as easily and unconcernedly as a human being could do. By a subdued grunt he answered to his name and to the whistle of his owner. Before crossing the road he would look to the right and then to the left to see if it was sufficiently clear of cabs and other vehicles for him to cross in safety. If so, he would scamper over to the opposite pavement, and then with a half toss of his head, and many wags of his curly tail, give a grunt of satisfaction at having performed the feat so cleverly. The pig had been taken home by his master when very young, fed by hand, allowed to run about the room like a child, and taught to follow him, etc., in the way we have described.

PIG FOUND IN A SEWER.—Many stories have been told of the large numbers of ferocious rats infesting the common sewers of London ; but, strange as they appear to be, the following one is even more so. We have been informed on good authority that about thirty years ago a sewer-man, in examining the sewer in the neighbourhood of Gray's Inn Road, found a black and white pig, of the Hertford breed, and about half-grown, fast asleep in a recess at the side. How it came there, or how long it had been in this dark subterraneous place, were matters which did not much concern the man at the moment. Seeing, however, a porky prize, and one well worth having, as the animal was in good condition, he forthwith seized it with a view to introduce it to far higher, lighter, and purer regions than it had been accustomed to for some time. The pig behaved with all seeming gravity when being carried along, and by passive submission and silence paid due respect to his captor, until they began to ascend the *man-hole*, when he commenced to kick and squall, as if a ring was being put through his cartilaginous extremity ; and it was not until after considerable difficulty that the sewer-man succeeded in landing his prize in open daylight, and fairly on *terra firma*. The pig appears to have survived this change in his condition from dark obscurity to light popularity, as he was exhibited, during six months, in a show by his captor, who charged so much admission for a sight of this sewer-prodigy, by which he made a great amount of money. How long the pig lived, or how he fared after absorbing so much public attention, and exciting so much curiosity about his antecedents, our informant knew not, but vouches for the truth of the tale as far as it goes.

'A HOG WITH A WOODEN LEG must be indeed,' as the writer of the following story says, 'a touching and picturesque object. It appears the cow-catcher at Dunleith took off the hind-leg of Mr. Smith's hog, and his benevolent owner made a wooden succedaneum and strapped it to the stump. The creature shows its gratitude by accumulating fat with great rapidity, the only drawback being that when killing and curing-time comes round, one of the hams will have no handle.'

CHAPTER XIV.

ARISTOCRACY OF ANIMALS.

Bold stag, with antlers large and strong,
Thou art a noble creature ;
And one more so we scarce could find
Throughout the world of nature.

IF all the various races of men constitute one universal brotherhood, and if ' a man's a man for a'
that,' there are nevertheless amongst them differences of size and complexion, and of habits
and life, which render them very distinguishable
one from the other. Even in the social position and mental
attainments of those who are of the same race and nation,
there are also wide differences.

Some are of noble descent, are wealthy, have refined
manners, move in the higher spheres of life, and have, in
many cases, great influence over their fellow-men ; and are,
on account of these things, considered to comprise what is
called the aristocracy of society.

The animals to which we are about to refer may, for more
reasons than one, be reckoned amongst the higher order, or
the aristocracy, of animals ; certainly more so than the pig,
and many other quadrupeds, whose habits of life are very
suggestive of the vulgar manners and vitiated tastes of many
who belong to the ' lower classes ' of human beings.

It was on a bright and genial morning in May, that

' Sweet month
If not the *first*, the *fairest* of the year,'

when white and pink may-blossoms adorned with beauty green hedgerows and hawthorn bushes, that we visited an extensive park in one of the western counties of England, which was skirted on one side with a plantation of larch, Scotch fir, and other trees. Fine old elms here and there, reared during the lapse of ages, towered upwards, and massive oaks spread out their gigantic limbs, affording a cooling shade to the cattle from the hot rays of the morning sun.

AN OLD MANSION.—In the distance stood a fine mansion alone in its own solitary grandeur, which, by its ancient style of architecture, seemed to point to mediæval times. Silently but unmistakably it told its tale of departed ancestral glory and greatness, and of the long since gone ones whose mortal remains rest in the mausoleum hard by, and at the same time it was painfully suggestive of the changes, the fleetness, and uncertainty of human life.

In the midst of these reflections, we were aroused by perceiving a numerous herd of deer rapidly coming towards us. Their movements indicated so definite a purpose, that we were at once seized with alarm. The deer were not more than a hundred yards off when a warning voice bade us make haste. We did so, and soon found ourselves within an enclosure, containing numerous sheds and a cottage, the residence of one of the park-keepers.

'You've had a narrow escape, sir,' he said, as we entered.

'Are the animals dangerous?' we inquired.

'Why, sir, not always,' he replied; 'but they are at certain seasons. You are, however, quite safe here, and I'll see they do you no harm when you leave.' He was as good as his word.

The incident referred to having produced in us a strong feeling of interest in the deer tribes, we shall now endeavour to describe their structure, habits, and uses.

DEER.—The deer family belong to Order IX. of the class Mammalia, usually named *Ruminantia*. The word is derived from *rumino*, from the Latin, *rumen*, the cud. Animals of this order when feeding swallow their food without masticating it, or but in a small degree; when at rest, they can bring it up again and grind it properly, which is termed 'chewing the cud.'

To all ruminant animals in their wild state this arrangement is indispensably necessary, because while feeding they might be preyed upon by their carnivorous enemies, which are very powerful and numerous. They are therefore provided with stomachs capable of receiving a large quantity of food in a crude state. This, as before intimated, they can re-chew when they find themselves in a place of safety. This peculiar organization appears all the more necessary and important, because comparatively there is but a small percentage of real nourishment to

Elk.

be obtained from the quantity of food taken; it is therefore obvious that the digestive apparatus should be of such a character as to animalize the food or to extract from it every particle of nourishment that it is capable of yielding.

We are told that the first stomach contains the food as it is first swallowed; the food then passes to the second stomach or honeycomb-bag, in which it is made up into balls, which are thence returned for re-mastication. After this the food passes to the third stomach, and then to the fourth, which is the next in size to the first stomach; this last is the chief organ of digestion.

Reindeer.

The *reindeer* is a native of Lapland, where it abounds, and is highly valued because of the services it gives to man, and the comfort it yields him both when living and dead. To deprive the inhabitants living in those cold snowy regions of the reindeer, would be a greater loss to them than the destruc-

tion of the palm and orange groves of Italy would be to the
natives of that sunny clime.

The reindeer draws the Laplander in his sledge over dreary
wastes of snow.   It gives him wholesome and nutritious milk,
of which he makes cheese ; and when dead its skin supplies
him with warm clothing, its flesh with food, and even its
tendons are used, when split, as threads, and when whole as
strings for his bows. If in England a man's riches are
reckoned by the number of pounds he may be worth, a Laplander's are valued by the number of reindeer he has in his
possession.

THE STAG, or male of the *red deer*, is a noble-looking
animal, and was at one time
much more numerous in this
country than it is now.   It
is common to all the northern
parts of Europe, and in the
Highlands of Scotland it is
found in great numbers in its
wild state.

' The red deer,' says Baird,
' is an ancient denizen of the
forests of this country, and
is intimately blended with the
old oppressive forest laws,
and with some of our legends
of deadly feud, such as "Chevy
Chase," etc. Its flesh is highly
esteemed as an article of

Stag, or Red Deer.

food.   In winter they go together in large herds.   The stag
sheds its horns annually, from the shavings of which we obtain
hartshorn.   He is very bold and courageous in time of danger,
and will defend himself in a very fierce and determined
manner.

We have heard of a tiger being put into an enclosure with
a stag, which made such a stout resistance that the tiger was
feign to fly.   A few weeks since, a stag was so closely pressed
by the huntsmen, that he turned at bay, closed in upon one
of his pursuers, and ran his antler into the chest of the horse,
which died a few minutes afterwards. This occurred in
Ireland.

AFFECTION OF THE STAG.—The stag is not only a bold and noble animal, but is very affectionate towards those of its own kind, and particularly when they are in pain or suffering. He has been known to lick the wounds of his companions that have been shot at but not killed, and in this way to administer relief.

Animals of the deer tribe seem to have a deep sense of any wrong done to any member belonging to them, and to show a determined disposition to avenge that wrong at the first opportunity. Captain Brown states that, ' At Wonersh, near Guildford, the seat of Lord Grantley, a fawn was drinking in the lake, when one of the swans suddenly flew upon it and pulled it into the water, where it held it under until it was drowned. This act of atrocity was noticed by the other deer in the park, and they took care to revenge it the first opportunity.

A few days after, this swan, happening to be on land, was surrounded and attacked by the whole herd, and presently killed. Before this time they were never known to attack the swans.'

Blending with the majestic appearance of the stag is a kind of plaintive, half melancholy, and pathetic look of his beautiful eyes. It has been said that he sheds tears. To this Shakespeare alludes where he says—

> ' The big round tears
> Coursed one another down his innocent nose
> In piteous chase.'

We are informed that this weeping aspect of the deer tribe ' depends on a remarkable glandular sinus, or tear-pit, situate at the inner angle of each eye, close to the nose, without having any communication with the eyes, or without what are termed the lachrymal passages.'

THE FALLOW DEER has long been domesticated in this country. It resembles the stag in many particulars, but is much less in size, and much more gentle. The horns incline slightly forwards, and have small branches behind them. It was originally brought from Persia, and is now found in every part of Europe. In England we have one variety of fallow-deer distinguished by a dappled skin, and another by a brownish-bay colour, more or less beautifully spotted, and by a tolerably long tail. This last variety was brought from Norway by King James I. The skin of the fallow-deer is remarkably soft and durable, and its flesh, called venison, is very rich and delicate. The female produces one, and sometimes two, young at a birth. This species go in large numbers, but often will, in the same park, be divided into two parties, which frequently contend, in thorough aggressive fashion, for some favourite or particular part of their domain or pasture-ground. Of course one of the males (for they are the males

Fallow-Deer.

that fight) becomes master of the field, and with his own party of females and subordinate males takes possession of the coveted part of the park. The vanquished one has also his party, to which he acts as leader. The leaders are usually, if not always the oldest, the strongest males of the two herds.

Fawns, the young of the fallow-deer, have a light, graceful, and beautiful appearance. They look like paragons of simplicity, innocence, and gentleness, and are, when kindly treated, so amiable and sociable in disposition, and so interesting in their little frolics, that we need not wonder that tender and loving children and humane people should become ardently attached to them.

A PET FAWN.—A very intimate friend of the author told him that she once took to her home a very young fawn that had been deprived of its maternal parent, wrapped it in flannel, put it in a basket by the fire, and carefully attended to its wants. This little animal soon became sufficiently familiar with its humane friend as to drink milk and to eat food out of her hand. It would frisk about as gaily as a lamb in a sunny field, and follow her from room to room, into the yard, garden, road, or wherever she went. When the lady was absent it would make a whining noise until she returned. When she did so it would show the most unbounded pleasure. As time passed on the fawn increased in size and strength, and when the weather became warm and genial it was turned into the park with the rest of the deer, amongst whom it no doubt felt perfectly happy.

The plaintive cry of the fawn is generally an indication of distress arising either from fear, pain, want, or a sense of the absence of the parent and the other deer ; and it seems to arouse, in an eminent degree, the affectionate concern of the older animals. We have been told that the cry of the roe when deprived of her young is one of extreme agony.

Deer are furnished with additional nostrils, or rather breathing-places, which are of great advantage, by giving them a fuller and freer respiration, specially needed by them in the chase. The additional nostrils are considerably above the ordinary ones, and are situated a little below the eyes. They are composed of folds of the skin which the animal can open and close at pleasure. This remarkable organization enables the deer to plunge their noses very deep into the water without suffering any inconvenience in breathing. This is a great boon to the animals, especially when the weather is hot, and when they have been made thirsty by the chase.

THE ROEBUCK.—The roebuck also belongs to the deer family. It is not more than about two feet high and three feet long, but it is an exceedingly graceful and nimble animal, and has been known to foil repeated efforts to capture it. In summer-time its coat is smooth and of a reddish colour, but in winter grey. The horns are much used for making knife-handles, etc. The horns of the roebuck differ from those of the stag and the fallow-deer, inasmuch as they have no basal snag, but rise straight from the forehead, throwing out one

antler in front and one or two behind.   The horn, from its
base, is covered with wrinkles.

These animals go in pairs only, and usually show great
affection for each other. They are specially tender and careful
over their young, for which they provide until they are old
and strong enough to do so for themselves.   They were at
one time common in our woods, but are now seldom seen
except in the Highlands of Scotland.   Some are found in
Italy, France, Sweden, Norway, and Siberia.   While fallow-

Roebuck.

deer seek the plains, the roebuck prefers higher grounds.   It
feeds upon the tender shoots of underwood as well as upon
herbage.   When properly killed its flesh is very delicate for
food.

DEER-STEALING.—When the forests in England were more
numerous and extensive than they are now, deer-stealing was
a very common offence, and became at last of such a daring
character that laws had to be passed to put a stop to it, or at
least to check it.   Any person found disguised and in arms
in any forest or park, and killing red or fallow deer, was
deemed a felon without benefit of the clergy.

It is recorded of Shakespeare that in early life he fell into bad company, amongst whom were some that followed deer-stealing, with whom he was more than once engaged in robbing Charlecote Park, near Stratford-on-Avon, belonging to Sir Thomas Lucy; for this he was prosecuted, and in order to be revenged wrote a ballad against that gentleman, which incensed him to such a degree as to cause him to commence a criminal prosecution against Shakespeare, who was in consequence compelled to flee to London for shelter and protection.

This apparent piece of ill-fortune appears to have been the first stepping-stone to that far-famed notoriety which Shakespeare afterwards attained as a writer. It is said that in the time of Queen Elizabeth it was a common practice for persons to ride to the London theatres on horseback. During the performance the horses of those who had no servants were usually given in charge of anyone who might be waiting for such employment. Shakespeare became a popular favourite as a caretaker of horses, then was introduced as a servant in one of the theatres; afterwards he became a performer, and finally one of the most powerful dramatic writers of his day.

A NIGHT'S SURPRISE.—Once upon a time in the autumnal season of the year we were walking by the side of an extensive park bounded by a high wall overhung with trees. It was late at night, the weather rather squally, and the rain, which had fallen during the day, had rendered the road so soft that we could hardly hear the sound of our own footfall. It was very dark. There was no moon, and the stars were hidden by inky clouds. Suddenly we heard a most distressing, half-moaning voice, which seemed to say, 'Oh dear! oh dear!' proceeding from some one apparently close by.

We immediately stopped and looked timidly round to ascertain where the strange noise came from. But there was no sign of life or movement to throw any light on the mystery. After a dead silence of a minute or so the same cry was repeated, rendering our perplexity and the mystery all the greater. We then hurried with rapid steps from the place, and in about half an hour were relieved by seeing some lights in the houses which here and there stood by the high-road side.

We told this little adventure to our host at whose house we were staying for a short time. With mischief lurking in his small but quick eyes he stated that just about the spot where we had heard the mysterious noise a man some years before had been found murdered, and that it was reported by superstitious people to be haunted by his ghost, where it had often been heard to utter a sound resembling 'Oh dear! oh dear!' as if in great agony.

'But, sir,' he said, 'your mystery is easily solved. I have no doubt that at the other side of the wall a herd of deer were sleeping, and, as they are very quick of hearing, you may have disturbed them when you were walking past. The noise which alarmed you was very likely the cry of a fawn which wanted its mother, and which cry sounds much like the utterance of "Oh dear! oh dear!"' Was not this a dark and deer adventure?

FREAK OF A STAG.—A correspondent of an *illustrated* paper gives the following interesting account of a very sociable deer. The writer was quietly seated in a railway carriage at Wimbledon Station, when a face appeared at the window, and, looking up, he saw that it was a stag. The animal regarded him placidly, and seemed inclined to enter the carriage, but on second thoughts turned away. There were a lot of people on the platform, and crowds outside the station shouting, but the stag did not seem at all startled ; on the contrary, he coolly walked off, and, seeing an opening between two carriages, walked up to it, contemplated it for a moment, evidently concluding that it was not a fair sporting fence. He then crossed the rails, and sprang on to the upplatform, where he gave a kick and a jump—or rather a jump and a kick—at a cabby who tried to stop him, bolted through the station, and up the bank, over which he disappeared ; but evidently having a fancy for the railway, he recrossed the line a little lower down, ran about amongst the fitters' and carpenters' shops, and was captured by some workmen. Two couple of hounds cast up just as the stag returned to the rails, and immediately afterwards the 2.30 West of England express rushed through the station, fortunately without doing any damage. That a stag should come upon a busy railway platform when a train was in the station seems to be curious.

CHAPTER XV.

AN ANCIENT FAMILY.

'That sheep-cot, which in yonder vale you see
  (Beset with groves, and those sweet springs hard by)
I rather would my palace wish to be
  Than any roof of proudest majesty.'—DANIEL.

HERE are some men who congratulate themselves on being able, by the aid of family records, to look back on a long line of predecessors, distinguished in their time for great wealth, influence, and philanthropy. The Jews are proud of tracing to a very remote period the origin of their race. The gipsies, in many cases, boast of being the descendants of the Shepherd Kings, whose invading proclivities were much dreaded by some of the monarchs of that distant period. Modern Rechabites claim the honour of being members of the same kind of order spoken of by one of the Old Testament prophets; and one religious denomination justifies its observance of the rite of baptism by the example of the forerunner of Jesus Christ.

Without intending to be at all invidious in identifying animals with human beings, we may observe that the ancestors of the subject of this chapter are mentioned in the Scriptures at a very early date. In Genesis iv. we read, 'And Abel was a keeper of *sheep*.' This fact is, we think, a good reason for selecting the title we have given—'An Ancient Family.'

We shall now give some interesting particulars of the structure, commercial value, and peculiar characteristics of this well-known and useful animal.

THE SHEEP FAMILY.—Sheep belong to the genus *Ovis*, and are a sub-family of the large family *Bovidæ*, or hollow-horned ruminants. There are very many varieties of them. Those in England are known as 'White Dorsets;' 'Old Norfolks;'

Sheep.

'South Downs;' 'New Leicesters;' 'Cheviots and black-faced Scots,' common in the North; and the 'Lincolns.' Small black-legged sheep abound in Wales, where they live on the scanty herbage growing on the sides of precipices, which are often so deep and look so dangerous that a person is apt to turn giddy by gazing at them. The smallest breed are the Shetlands, and their wool is the finest produced in Great Britain.

These different breeds may be distinguished by slight peculiarities of size, colour, and the length and quality of their wool, but they do in reality merge into one. There is some difficulty in tracing the origin of our sheep, and the country from which it was imported. Some naturalists believe the *moufflon*, an animal about the size of a small fallow-deer, inhabiting parts of Corsica, Sardinia, and Greece, to be the progenitor of the common sheep.

Moufflon.

The precise period of its introduction into England is also involved in obscurity. We are told, however, that the ancient Britons used to wear felted cloth more than 2,000 years ago. The sheep, being a ruminant animal, chews the cud the same as the ox. It has no teeth in the upper jaw, but has, in their stead, a hard, firm, but somewhat elastic pad, between which and the teeth of the lower jaw the herbage is secured, and this, by a sudden jerk, half-biting and half-tearing action, is separated from the stalks and roots and passed into the stomach, in which it remains until remastication takes place.

If the sheep is defective in courage, its social instinct is remarkably strong. They feed and travel together in flocks if left to themselves, and should one of their number jump a fence or find its way through a gap in the hedge into another field, the rest will be sure to follow.

It is no doubt on account of this faculty of imitation in sheep that butchers will sometimes lay hold on one of a number of them, and force it into the building or slaughter-house he may desire, knowing that the others will follow.

This plan is usually successful. There are but few animals that appear to be more happy and contented when living together in numbers than sheep; and yet so great is their dread of isolation, that if one of them is separated from its companions it refuses food, pines, and very soon dies.

MOTHER SHEEP AND THEIR YOUNG.—Innocence and implicit obedience to the will of the shepherd are commendable traits in the character of these animals. They are also remarkably affectionate to their young. Hogg, the Ettrick shepherd, says : ' The harder the times, the greater the kindness of the ewe to her young. Once I herded for two years in a wild and bare farm, called " Willenslee," on the border of Midlothian ; and of all the sheep I ever saw, these were the kindest and most affectionate to their young. We had one very bad winter, so that our sheep grew lean in the spring, and disease came in among them and carried off many. Often have I seen these victims, when fallen down to rise no more, and even when unable to lift their heads from the ground, holding up the leg to invite their starving lambs to the miserable pittance that the udder could still supply.'

If a mother sheep lose her lamb, what a bleat of desolation she will give ! and how, with all the affectionate concern of a mother, will she look and run hither and thither among those of her own kind until she finds the lost one. This maternal feeling is, however, exhibited in a more prominent and affecting manner when her young one dies. If it is not removed she will cling to its decaying remains until they disappear by mixing with the soil.

Not only is the flesh of the sheep very nutritious and commonly eaten in this and other temperate and northern countries, but these animals constitute, in various ways, very profitable speculations in our commercial enterprise. It is said that a few years since the flocks of sheep in several of the European kingdoms numbered 140,000,000 head.

An observer of the habits of sheep has informed us that these animals usually begin to feed in the summer-time about five or six o'clock in the morning, but if it is likely to rain soon after this time they will not stir, but remain in the same places they have occupied during the night, simply because if they did not do so, and the rain were to fall while they were feeding, they would have no dry places to lie down upon.

Although this animal is much exposed to cold and wet weather, it is provided with protection against both. We must condemn, as a cruel act, the custom of shearing sheep too early in the spring, which often brings on suffering, and deteriorates the value of the animals.

THE WOOL OF SHEEP.—It would be difficult to ascertain the vast amount of money paid in combing, dressing, spinning, and dyeing the wool of sheep, and then weaving it into cloth afterwards, to say nothing about the cost of machinery and looms, etc., etc. In addition to this, we may see in our large manufacturing towns, as well as in those of other countries, thousands of men, women, and children employed, and deriving their livelihood by working in some way or other on the wool of these useful animals.

If we examine a single hair of sheep's wool through a powerful lens, we shall discover a number of serrations, which resolve themselves into a series of notched ridges, which surround the hair closely. By means of these serrations the hairs interlock with each other in what is called 'felting,' which is increased by water. It is this that renders it necessary to have cloth well shrunk before making it up into garments, coats, or trousers, as the first drenching by rain will make them too small for the wearer. The reason this shrinking does not take place when the wool is soaked with water while on the sheep's back is, 'that the fleece is imbued with a secretion from the skin, called the "yoke," which repels the action of the water.'

From what we have stated, it will be seen that if sheep lack the size and strength of oxen, and the fleetness of horses, and if they are less bold, intelligent, and courageous than many other animals, they are not, on the whole, less useful, but have in many cases some claim on priority in contributing so largely to the general comfort and convenience of mankind.

CURIOUS USE OF SHEEP.—'In the "Colonies and India,"' says *Nature*, 'we find a note respecting the employment of sheep as beasts of burden. In eastern Turkistan and Thibet, for instance, borax is borne on the backs of sheep over the mountains of Seh, Kangra, and Rampur, on the Sutlej. Borax is found at Rudok, in Changthan, of such excellent quality, that only 25 per cent. is lost in the process of refining. The Rudok borax is carried on sheep to Rampur, which travel at the rate of two miles a day; but notwithstanding the superior quality and the demand for it in Europe, the expenses attending its transport seriously hamper the trade, which, but for the sheep, would hardly exist at all.

A WOOLLY AUDIENCE.—Six hundred years ago a certain shepherd boy watched his flock on the hills that look down upon Florence, and spent his leisure in piping to the sheep, and making sketches of a favourite ewe or lamb 'with a stone slightly pointed upon a smooth, clean piece of rock.' This little boy, Giotto Bondone, became one of the greatest of Italian painters, and ever took delight in portraying the gentle creatures in whose company he had passed his boyhood. Sheep are very fond of music. 'Joseph Haydn, when a boy, went on a tour with some companions through the Apennines. One of the party carried his flute with him, and one day, as he sat on a hillside and played for the amusement of the others, the sheep came crowding round him. If the time was slow and mournful, the sheep would droop their heads; but when he played a lively strain, they drew close to his side, and rubbed their necks against his legs to show their delight.'

THE PHILOSOPHER AND SHEPHERD BOY.—According to a story told of Sir Isaac Newton and a shepherd boy, sheep are good indicators of coming weather. It appears that the great philosopher was on one occasion riding across Salisbury Plain, where he met with a boy tending sheep. Whether there was any sign or not of rain we cannot say, but Sir Isaac, after sundry interrogations, asked the boy 'what he thought of the weather?' To which the boy replied 'it was going to rain,' and advised the traveller not to proceed on his journey. His advice was unheeded. The philosopher had not proceeded far on his journey before the boy's answer occurred with great force to his mind, and so curious did he become to know *why* the boy should give him such a reply, that he resolved to retrace his steps and ask him. He did so.

'How do you *know* it is going to rain?' he asked the boy; 'and what is it makes you so certain about it?'

'Just this,' answered the unsophisticated lad, pointing to an old ram; 'whenever he turns his tail towards the wind, I *knows* it will rain *afore* long. It al'a's does, sir, and that be true, sir; and his tail has been in that direction pretty nigh all the morning, it has, sir.'

We do not know whether the philosopher's faith was sufficiently great to believe in the barometrical qualifications of the old ram, or in the assertions of this untutored lad, to induce him to give up his journey or not, but he no doubt thought it was a subject worthy of his attention, study, and reflection.

Sheep have been known repeatedly to seek the shelter of a hedge, an embankment, and the side of a hill, a considerable time before a storm or rain has come on, and even when there has been no indication of either.

A SHEEP'S TASTE FOR MUSIC.—The Rev. T. Jackson says, in referring to sheep being fond of, and variously affected by, music, that the Highland breed of sheep carry off the palm for cleverness and for their partiality to sweet sounds. He knew one of them that would jump and skip about with considerable pleasure whenever a lively, quick tune was played; but the moment it heard the National Anthem, it would hang down its head, appear to be very sullen, annoyed, and much displeased until the music ceased.

A TIPPLING LAMB.—There can be no doubt that when animals are not controlled by man, they seldom, if ever, acquire a fondness for anything in the shape of food or drink but what they are prompted to take by the law of instinct. We, however, once knew a lamb whose parents died prematurely, that was taken home by its owner, who kept a wayside inn in the county of Gloucester. This lamb was fed with milk from the cow, on which it thrived amazingly well. As it grew older, it became very familiar with the domestics, as well as with the customers in the tap-room. Keeping this kind of company, singular as it may appear, it acquired a taste for malt liquors, of which, the landlord told us, it had a glass regularly every day, and that immediately afterwards it would frolic and jump about in the most lively manner, showing that beer had the same effect on the brain of this animal as it usually has on those of human beings.

ODD COMPANIONS.—In 'Records of Animal Sagacity' appears the following account of a singular attachment which two animals of different tribes formed for each other. It seems that 'In December, 1825, Thomas Ray, blacksmith, Handhills, parish of Brittle, purchased a lamb of the black-faced breed from an individual passing with a large flock. It was so extremely wild, that it was with great difficulty separated from its fleecy companions. He put it into his field in company with a cow and a little white galloway. It never seemed to mind the cow, but soon exhibited manifest indications of fondness for the pony, which, not insensible to such tender approaches, amply demonstrated the attachment to be reciprocal. They were now to be seen in company in all circumstances, whether the pony was used for riding or drawing. . . . When likely to be too closely beset, the lamb would seek an asylum beneath the pony, and pop out its head betwixt the fore or hind legs, with looks of conscious security. At night it invariably repaired to the stable, and reposed under the manger, before the head of its favourite. When separate, the lamb would raise the most plaintive bleatings, and the pony a responsive neighing. On one occasion they both strayed into an adjoining field, in which was a flock of sheep; the lamb joined the flock at a short distance from the pony, but as soon as the owner removed him, it quickly followed, without the least regard to its own species.'

A SHEEP THAT CHEWED TOBACCO.—In volume xxii., and on page 329, of *Nature* appears the following, which we copy, as being something very curious and uncommon :

'The subject of a depraved taste in animals is an interesting one, which has not been studied as much, perhaps, as it might. In human beings it would seem to depend on ill-health of either body or mind; but in animals it would seem as if it might be present and the animal enjoy good health. One remarkable instance in an herbivorous animal we can vouch for. It occurred in a sheep that had been shipped on board one of the P. and O. steamers to help to supply the kitchen on board; but while fattening, it developed an inordinate taste for tobacco, which it would eat in any quantity that was given to it. It did not much care for cigars, and altogether objected to burnt ends; but it would greedily devour the half-chewed quid of a sailor, or a handfull of roll tobacco. While chewing, there was apparently no undue flow of saliva, and its taste was so peculiar, that most of the passengers on board amused themselves by feeding it, to see for themselves if it were really so. As a consequence, though in fair condition, the cook was afraid to kill the sheep, believing that the mutton would have a flavour of tobacco.'

CHAPTER XVI.

LOWINGS FROM THE FIELD AND SHED.

The bellowing ox and lowing kine
  Are treasures worth e'en a golden mine ;
These pictures of life, in fields of green,
  Can vie with the richest ever seen.

MERE cursory view of the life of various kinds of animals may lead us to suppose that there is a very wide difference in their utility to man. Admitting that it may be so to a certain degree, yet if we look carefully into their positions in nature, and the habits and services of each tribe, we shall find the difference referred to is not so great as we may imagine. One kind of animal may in some particular thing be much more useful than another kind, but at the same time may be very inferior to the latter in something else.

If the horse is the fleetest, the strongest, and most useful of our domestic animals as a worker, he produces no material to compete with the wool of the sheep, neither is he of such use to man after death as the latter animal is. If cows are not so useful as the horse as workers, or if they are not wool-producers like the sheep, yet in the milk they give we have an incalculable blessing enjoyed by man in nearly every part of the world. We should look at the nature or quality, as well as to the extent, of the advantages we derive from animals, in order to form a proper estimate of their real value to man.

We shall now give a brief sketch of

THE OX AND COW.—These animals belong to the order *Ruminantia*, class *Mammalia*. Many varieties of them are found in Europe, India, Africa, and in North America, some

of them domesticated, but most of them living a wild life.
The species belonging to England are characterized by flat
foreheads, wide muzzles, heavy bodies, and strong legs. The
horns of some of them are long and curved, others are short,
and a few are without them; but *all* these animals contribute
in an eminent degree to the wants and comfort of man.

What sight can be more beautiful and interesting, or more
suggestive of prosperity and enjoyment, than herds of these
noble creatures grazing on the mountain-side, or in verdant
fields, where cowslips and buttercups gleam in the sunlight
like vast sheets of the purest gold? We in England may well
be proud of our oxen, and should be grateful for them too,
because they not only give life and animation to our landscapes, but they afford to man a number of blessings he
would have to look for in vain in many other animals.

Let us stand in imagination for a short time in a field where
a number of them are feeding, and watch their movements;
we shall find them peculiar, interesting, and amusing, and we
may learn something at the same time.

STRUCTURE OF THE COW'S MOUTH.—We shall see that
these animals, when grazing, throw their tongues round several
blades of grass or herbage, which they gather together in the
same manner as the reaper does stalks of corn by his sickle;
they then grip them firmly between the teeth of the lower
jaw and the pad of the upper one; then give a sudden jerk,
which separates the stalks from the roots.

The reason why cows and oxen use their tongues in this
manner, is that their upper lip is not prehensile like that of
the horse. Their inability to seize anything by their upper
lip is therefore supplied by the movement of their tongues as
before described. As they require a great deal of food, it
takes a long time before they obtain a sufficient supply; and
as they are during this time exposed to the hot sun, and to a
thousand winged tormentors, nature has generously furnished
them with long tails to ward them off, or at any rate to lessen
the misery which they would otherwise have to endure if it
were not for this useful appendage.

ORIGIN OF OUR OXEN.—Naturalists are divided in opinion
as to the origin of our oxen, and the time of their introduction into Great Britain. Some think they came from
some part of Asia, and that the Romans brought them hither.

Those who believe they are derived from wild oxen, which thousands of years since existed in England, have in their favour the fact that fossil bones, resembling the bones of our domestic ox, have been found in our rocks and caves.

WILD OXEN.—The 'Student's Natural History' states: 'The cattle which anciently inhabited the great Caledonian forest, roaming wild and free, was a small kind, and the breed is still preserved, though they are restricted now to a very few places, such as Cadzow Park, near Hamilton, and Chillingham Park, in Northumberland.'

In order to show the superior power and influence of man over the lower animals, and the wide difference which exists between animals of the same genus in a wild and domesticated state, a description of the herds referred to may be both interesting and useful. It is given by a gentleman named Culley, who visited Chillingham Park some years since, and who describes them in the following language : 'Their colour is of a creamy white, muzzle black. At the first appearance of any person, they set off in full gallop, and at the distance of two or three hundred yards make a wheel round and come boldly up, tossing their heads in a menacing manner. On a sudden they make a full stop, at the distance of forty or fifty yards, looking wildly at the object of their surprise; but upon the least motion being made, they all again turn round and fly off with equal speed, but not to the same distance. Forming a shorter circle, and again returning with a bolder and more threatening aspect than before, they approach much nearer, probably within thirty yards, when they make another stand and again fly off. This they do several times, shortening their distance and advancing nearer, till they come within ten yards, when most people think it prudent to leave them, for there is little doubt but in two or three turns more they would make an attack.' These animals, like many other wild ones, gore to death those that are wounded, feeble, or aged among them.

Admitting that the progenitors of our present race of cows and oxen were wild as those described, we cannot fail to see what a subduing and controlling power man has over brute beasts, and how willingly subservient these creatures become to his will, especially when treated kindly and humanely by him.

14

VALUE OF COWS.—In whatever country these animals are found, they are pre-eminently useful to the inhabitants of it, as the following calculation will show. Taking for granted that on an average cows produce 24 pints of milk per day each, that would amount to 168 pints per week. Supposing that 3,000,000 people in London drink each a pint of milk per week, then 17,857 cows are required to supply this quantity, which at twopence per pint would amount to £25,000 weekly. If the above number of cows, after giving milk for some years, be sold to butchers at £15 each, they would realize in money value £267,855.

Assuming the number of cows in England to be only

Yorkshire Cow.

178,570, these sold at £15 each would realize £2,678,550. If to these milk-producers we add the same number of oxen at the same price, then we have in these animals together the worth of £5,357,100. We must also consider that the amounts stated are greatly increased by the retail prices paid by con- sumers of the flesh of these animals, and also by the sale of calves, of which many thousands are slaughtered in one year.

In addition to the flesh of these animals, other parts of their bodies are of great service to man. Out of their horns drink- ing-vessels are made; glue, so useful to the mechanic, is made from their feet and ears; their hides, tanned into leather, give employment to thousands of men, and, when cut up into shoes and boots, protect the feet of millions of people; and their hair is mixed with mortar for the ceilings of our rooms; so that, living or dead, they may be regarded as the most useful animals that were ever brought under the control of man.

COWS AND CANDLES.—Before gas, and many other artificial lights now in use, were introduced, the common tallow candles were much more valued than they are at the present time. Although the poorer class of our ancestors were compelled to pass the long and dreary winter evenings comparatively in semi-darkness, the old-fashioned candles were nevertheless regarded as a great boon.

The following curious statements, which appeared in a London newspaper, may not only surprise the reader, but will, we venture to think, in a great measure enhance, in his estimation, the value of the ox and cow, whose fat, as well as that of similar animals, constitutes the principal ingredient in the manufacture of candles. It appears that, some time since, 240 tons of candles were sent out from a factory in Lambeth. This number of tons represents about 5,000,000 of candles, reckoning a little more than eight to the pound weight. The paragraph referred to goes on to say that: 'If these candles had to burn one at a time, they would last 3,368 years; or, if the first candle had been lighted when Saul ascended the Jewish throne, and had been burning, one at a time, ever since, there would be a sufficient supply left for 112 years to come.'

Even admitting that this calculation is not strictly, or absolutely, accurate, it is a near approximation to the truth, and shows how much, even now, we owe to the ox and cow for contributing in so great a degree to what may still be regarded as the blessing of candle-light.

A COW THAT CAME TO THE RESCUE.—We remember reading a simple story of a cow that was grazing in a field with a number of sheep, one of which was lying upon its back, unable to turn upon its feet. It was surrounded by several of its companions, who, however much they pitied its helpless condition, seemed quite unable to render the assistance it required, or perhaps lacked the sense to do so. Their bleatings of regret arrested the attention of the cow, who, looking up, comprehended the difficulty they were in, pushed forward to the struggling sheep, put her horns gently under its body, and with an upward motion brought the animal again on its feet. The rescued one and the rest of the sheep bleated their best thanks to the cow, which she acknowledged with sundry switches of her tail, and forthwith resumed her feeding.

The affection of the cow for her young is at first exceedingly strong, but not so enduring as that of many other animals. When the calf is taken away her bellowings are very mournful, and even distressing, but they soon cease. Hence, no doubt, the saying, 'The bellowing cow the soonest forgets her calf.'

THE BUTCHER AND THE COW'S TEETH.—Referring again to the structure of the cow's mouth, the following incident may not be uninteresting. Some time since we were passing a butcher's shop in London, at the door of which stood a young man, of whom we made some inquiries relating to cows. 'Are you a butcher?' we asked. 'Yes, sir,' was the reply: 'I've been one about eleven years.' 'Have you ever helped to slaughter any cows or oxen?' 'Oh yes; hundreds of them.' 'Then you are well acquainted with the internal as well as the external structure of these animals?' 'No man more so, I should think.' 'Do you know if the cow has as many teeth in the front of her upper jaw as she has in her lower one, or if there is any difference?' 'To be sure I do,' he said, smiling at the remark. He, however, hesitated in giving us the information we wanted. He said at last, however, 'There is, I believe, about the same number.' 'Indeed!' we said; 'are you not aware that the cow has no teeth at all in the front of the upper jaw?' 'Nonsense,' said he; 'you are joking, or speaking of some old cow that has lost her teeth.' Calling his attention to a cow's head hanging close by, we said, 'Open the mouth of it, and let it speak for itself.' He did so, and his own too; and then exclaimed with great astonishment, 'Well, I never knew such a thing in all my born days. I always thought that cows had teeth at the top and bottom the same as I have.'

A man in Gravesend assured us that although he had been a cowherd thirty years he never knew before that the front of the cow's upper jaw was deficient of teeth.

The above shows that we may be in the daily habit of

looking at things, and yet not see them so as to understand what they really are.

Do Oxen Kneel on Christmas Eve?—We will now quote an extract from the ' Book of Christmas,' by T. K. Hervey, referring to a notion that ' oxen kneel on Christmas Day.' ' In the south-west of England there exists a superstitious notion that the oxen are to be found kneeling in their stalls at midnight of this vigil, as if in adoration of the Nativity : an idea which Brand, no doubt correctly, supposes to have originated from the representation by early painters of the event itself.' That writer mentions a Cornish peasant, who told him (1790) of his having, with some others, watched several oxen in their stalls on the eve of old Christmas Day. At twelve o'clock at night they observed the two oldest oxen fall upon their knees, and, as he expressed it in the idiom of the country, ' make a cruel moan like Christian creatures.' To those who regard the analogies of the human mind, who mark the progress of tradition, who study the diffusion of certain fancies, and their influence upon mankind, an anecdote related by Mr. Howison in his ' Sketches of Upper Canada ' is full of comparative interest. He mentions meeting an Indian at midnight, creeping cautiously along, in the stillness of a beautiful Christmas Eve. The Indian made signals to him to be silent, and when questioned as to his reason, replied : ' We watch to see the deer kneel ; this is Christmas night, and all the deer fall upon their knees to the Great Spirit, and look up.'

The Cow and Moral Suasion.—Cows, like other animals, have their tempers, which they sometimes exhibit when about to be milked. On these occasions we have seen them peremptorily refuse to have the shackles put on their hind-legs, which in many cases is done to prevent them from walking and from lifting up either leg, by which they might upset the milk-pail. A good cudgelling with the milk-stool, which cows often get for not quietly submitting to be thus manacled, has never, that we can remember, produced any beneficial results, but has, in many cases, increased the animal's objection to the shackles referred to. From the following anecdote it is evident that, obtuse as the intellect of cows is considered to be, these animals are susceptible to gentle treatment, which, as a rule, proves far more efficacious in soothing any irritability of temper than rough usage could possibly do. Mr. E. Powell writes suggestively to the *Rural New Yorker* of a 'spoiled' cow reformed by moral suasion :

'Coming home to farm on a vacation, after nearly a year's absence, my "man" said to me, "You will have to sell that pet Ayrshire heifer. She is a terrible creature; we can do nothing with her." At milking-time I found they had her with a rope around her horns, in between close bars, and then stout pins before and behind her hind-legs. One switched off flies, while another milked. She made it lively, however, in spite of bars, pins, pegs, ropes, and men. It looked like a bad case. Her eyes were full of mischief. The next day I had her led out on the lawn of nice grass for a good meal. The next day she was let loose in the yard, when I took a rope to the gate, held it up, and caused her to come and let me put it on, promising her a good time on the lawn. It was at least a half-hour's work of quiet, persistent waiting and talking. But she was slowly coming to the point of yielding. At last she held her head quietly down close in front of me, and not till then did I yield one inch to her. She must submit before she could have the coveted grass. Then I led her out at once and gave her a good time. The result was that in three weeks she could be milked anywhere on the lawn without the least danger.'

CHAPTER XVII.

FOUR-FOOTED HYBRIDS, OR HALF-AND-HALF RELATIONS.

A Mule art thou ; in breed not pure,
But hybrid in thy nature ;
And yet good qualities we see,
Of horse and ass, unite in thee ;
Thou hardy, useful creature.

THERE are, we should imagine, but few readers advanced in life who cannot call to mind the time when, as boys or girls, they derived almost infinite pleasure from old Æsop's most wonderful and interesting fables, as well as in the useful morals they were intended to teach. It may not be an inappropriate introduction to the subject of this chapter, if we refer to one of these fables, entitled 'The Boasting Mule,' and then, after giving a description of the temper, habits, and uses of this hybrid animal, leave our readers to form their own opinion of his merits, or demerits, as the case may be. It is said : 'There was once a favourite mule who was constantly boasting of his family and ancestors. "My father," said he, "was a famous courser, and I myself take after him." He had no sooner spoken the words than he was put to the trial of his speed, and proved a failure. At that moment, also, his father began braying, which at once betrayed the secret of his descent ; and the whole field made sport of the *boaster* when they found he was only the son of an ass.'

CRUELTY TO DONKEYS.—That the estimate formed of any tribe of animals in any particular country is no test of their intrinsic value and importance, may be seen in our remarks, in the eighteenth chapter, on the donkey. As we have there

shown, its lot in England is often of the lowest and most degrading kind; and that, in the majority of cases, he is the
abused slave of some of the most ignorant, mean-spirited,
selfish, and brutish men that ever lived, who know not his
worth, and who care less for his claims on proper treatment.
And yet we find that the same animal in ancient times, and in
Eastern countries, has always been held in the highest estimation, much valued for his utility, honoured with the most noble
and aristocratic associations, and favoured with the most dignified employment.

The same remarks, in many respects, may apply to the half
relative of the ass, the hybrid animal who forms the subject
of this chapter. The mule has been rendered conspicuous by
the absence of reference to him in many works on natural
history. Whatever reason may be assigned for this, whether
it be the paucity of the numbers of this animal, or that he is
not of pure descent, and that his associations are therefore
considered to be objectionable, we are not prepared to say.
We have, however, one of the best of all precedents for intro-
ducing the mule to the notice of the reader. Like the ass, he
is spoken of in the Scriptures, and identified with persons and
circumstances which stamp him as an animal highly valued
and esteemed during the prophetic and early ages of the world.
He is mentioned in the Book of Genesis. King David and
his nobles rode upon mules; and it is said these animals are
mentioned by the early heathen writers; that in the time of
Homer mules ploughed the plains of Greece; that they were
often employed in the chariot-race; that they dragged the
combustibles to the funeral pile of Patroclus, and the chariot
of Priam to the tent of Achilles.

We shall now give a description of the structure and general
character of this hybrid animal.

THE MULE AND JENNET.—The *mule* is the offspring of two
animals of distinct tribes, the male ass and the mare, or female
horse. The *jennet*, of the horse and female ass. Although they
are nearly related, they differ considerably in size and strength,
and therefore, in many respects, in utility. The jennet is much
smaller than the mule, but stronger and more enduring than
the pony. The mule is considerably larger than the jennet, is
very hardy, and, when well treated, may live to be forty years
old. The natural disposition of the mule may be inferior to

the horse, but he is not less useful. In mountainous districts he is highly valued because of his sure-footedness, his strong constitution, and great physical power, which enable him to carry heavy loads and to bear with safety ladies and others through dangerous Alpine passes. Mules, because of their hardy nature, rendered essential service during the Crimean war, as they survived the hardships and the cold which caused the death of very many horses.

'Very fine mules were formerly imported into England for the use of prelates of the Church of Rome.'

The Mule.

We are informed that in Spain mules are numerous, and are often employed to draw persons of very great distinction. Great care is taken to improve the breed of them; they are, therefore, very fine, beautiful animals, and frequently sell for fifty, seventy, and even one hundred pounds each.

JUDGES ON MULES.—Pennant says that in old times the judges rode to court on mules, but in the reign of Queen Mary they changed those restive animals for easy pads, a justice of the Court of Queen's Bench, first setting the example.

The mule has been accused of possessing a bad temper, of being obstinate and self-willed, of having a strong inclination to do as he pleases, without regard to words, to menaces, or to blows, and that if he condescends to do a thing it is only when *he thinks* proper to do it that it is done. Much of this may arise from ill-usage. We cannot, therefore, altogether condemn him under these circumstances. The results referred to are very similar to those that would follow unkind treatment of a human being, because, after all, human nature and animal nature are very much alike in this respect.

There can be no doubt that kind treatment, with proper care and attention to the breeding of mules, tend very materially to increase and to improve their physical power, general appearance, and utility.

MULE SHOW.—The following account of a mule show, held at the Crystal Palace in May, 1875, which appeared in the *Standard* newspaper, will be read with interest, if not with surprise :—' It is probably within the mark to say that many English farmers never have seen a mule, while the acquaintance of those who do not come within this category is probably confined to having seen the diminutive animals, generally crosses with a pony, which little exceed in size their immediate asinine progenitors. The simple statement that in Class First of the present show, for mules of fourteen hands and upwards, for farming and heavy draught work, the six entries are animals varying from fifteen hands three inches to seventeen hands high, and of the weight of average farm horses, ought to exercise the agricultural mind considerably, when it is considered in relation to the undoubted facts that such mules, in proportion to their weight, are much stronger than horses ; that they will live and thrive on food that no horse would look at ; that they are rarely sick, are vastly more intelligent than the horse, and do good service up to forty years of age. Mr. C. L. S. exhibits four of these splendid brutes, and Mr. C. A. R. H. two, all imported from Poitou, and, of course, all picked specimens, costing there from eighty pounds to one hundred pounds each.

' In this district of France mule-breeding has been for generations a *spécialité*, four hundred pounds, and even five hundred pounds, being asked by the owners for the powerful jackasses used for stud purposes. The department of Deux-Sèvres has long enjoyed a high reputation for its large mules, many of the fine animals used in Spain and Italy, for carriages, coming from this district ; while La Vendée and Charente furnish the powerful, sure-footed pack animals used for transporting merchandise over the steep passes of the Alps and the Pyrenees.'

It would appear from the above description, that not only to the farmer, in agricultural work, but in field artillery, mules would be a very useful acquisition, because their capacity for enduring exposure and bad feeding is much greater than that of horses.

THE MULE'S POWER OF ENDURANCE.—'Some idea of the mule's capacity for work may be gathered from a statement annexed to the winner of the second prize, Mr. J. C.'s " Polly," which is thirteen years old. In August, 1870, Polly, bred between a common donkey and an Exmoor pony, was driven from Weargifford, in North Devon, to Berkeley Square, a distance of two hundred and twenty miles, in forty-two hours, including all stoppages. In the first day one hundred and thirty miles were covered, and the remaining ninety miles in part of the next day, the last ten miles being done in an hour. This same little animal, with her sister Betsy, two years younger, which is exhibited with her, made a journey of two hundred and twenty miles in forty-eight hours, in April, 1873, conveying four persons with luggage.'

' The mule,' says Brown, 'possesses some of the best qualities of the two useful animals from which it springs. It is, indeed, inferior to the horse in strength, and to the ass in patience, but it retains somewhat of the agility and beauty of motion which we admire in the one, and is sure-footed like the other. It has a spirited look like the horse—it toughly endures labour like the ass; the external resemblance to both its parents is wonderfully preserved throughout every part of its body.'

Instances of the great intelligence of mules have been given by travellers on the Continent, especially in Spain, and we may add in Egypt, particularly in Cairo, where, as well as donkeys, they are very numerous, and may be seen standing in the squares for hire. Not only will they answer to their own names, and obey the word of command when being driven, but in Spain, when two are yoked together, if the words, 'Aquella otra' (meaning, 'You other also') are used, they know the words equally apply to both, and therefore they have the effect of stimulating them to renewed exertions and additional speed.

On the levee in New Orleans, harnessed in drays, mules may be seen who understand the French, English, Spanish, and German languages.

FEROCIOUS COURAGE OF A MULE.—Mr. Arnauld, in his 'History of Animals,' relates the following incident of ferocious courage in a mule : 'This animal belonged to a gentleman in Florence, and became so vicious and refractory that he not only refused to submit to any kind of labour, but actually attacked with his heels and teeth those who attempted to compel him. Wearied with such conduct, his master resolved to make away with him by exposing him to the wild beasts in the menagerie of the Grand Duke. For this purpose he was first placed in the dens of the hyenas and tigers, all of whom he would have soon destroyed had he not been speedily removed. At last he was handed over to the lion ; but the mule, instead of exhibiting any symptoms of alarm, quietly receded to a corner, keeping his front opposed to his adversary. Once planted in the corner he resolutely kept his place, eyeing every movement of the lion, which was preparing to spring upon him. The lion, however, perceiving the difficulty of an attack, practised all his wiles to throw the mule off his guard, but in vain. At length the latter, perceiving an opportunity, made a sudden rush upon the lion, and in an instant broke several of his teeth by the stroke of his fore-feet. The "king of the animals," as he had been called, finding that he had got quite enough of the combat, slunk grumbling to his cage, and left the hardy mule master of the battle.'

The author of a work, 'The Passions of Animals,' cites some very curious instances in proof of the assertion that the lower animals do, in many cases, use their experience in reference to things from which they have suffered pain and annoyance. He says, 'The dog which has been punished once for a fault will either slink away or hide itself if it finds itself detected in the repetition of it. The well-known story, recorded by Plutarch, proves the application of accidentally acquired experience. He says that,

' "A MULE LADEN WITH SALT fell into a stream, and having perceived that its load became thereby sensibly lightened, adopted the same contrivance afterwards purposely, and that to cure it of the trick its panniers were filled with sponge, under which, when fully saturated, it could barely stagger." '

CHAPTER XVIII.

OUR DONKEYS AND THEIR KINDRED.

Though thou art but an ass, we plainly see
   That something good is really found in thee ;
For thou art patient, sometimes even brave,
   A cheap, contented, uncomplaining slave.

HAT the same object is differently estimated and appreciated by different people is as true as that the hills are old. In nothing is this more clearly seen than in the fact that the same animal which is disliked and badly treated by one person, may be highly valued, well looked after, and admired by another. We were forcibly struck with the truth of what we have just said when passing one day through a West-end part of London.

It was early morning time. Many dealers in vegetables were returning with laden barrows and carts from Covent Garden Market. Amongst them was a repulsive-looking fellow, whose donkey was very poor, and, as Waterton says, looked like ' Misery steeped in vinegar,' and which at intervals received a curse, or a whack from a heavy stick, because his speed of locomotion did not please his brutal driver. ' Poor Neddy,' said we, ' yours is a hard fate.'

Soon after this we encountered another man and his donkey, who presented, in all respects, a widely different aspect. The man's face beamed with good-nature, kindness, and intelligence ; his donkey was in first-rate condition, and appeared to enjoy a happy existence, and as it stepped cheerfully along received, not curses and blows, but a few encouraging words.

What we have described gave rise to a train of thought as to the real cause of the difference in the lot of these two donkeys ; and we wondered whether it was traceable to the men or to the animals. That there was a cause for it was certain. We therefore resolved to study, more than we had done, the habits and qualities of the donkey family, so that we might arrive, if possible, at a right solution of this some-what difficult problem. We will now refer to the history, structure, and uses of

THE ASS.—This animal, though common, is not, as he ought to be, properly understood. Ignorant people have attri-buted to him a number of bad qualities he does not inherently

The Ass.

possess. No domestic animal in England has been more ill-used than the donkey; and yet we have no animal, taken altogether, that is of more real service to man, especially the poor one, than he is, or might be if he had his rights, and humane treatment extended to him, which, we may assert, would produce in this animal the most beneficial results.

The ass belongs to the order *Solidungula*, of the class *Mam-malia*. An animal belonging to this order has hoofs which are entire—that is, externally whole, and not divided in two like the hoofs of the ox, the sheep, and the pig. The word 'Soli-dungula' is derived from the Latin *solidus*, solid, and *ungula*, a hoof. To the same order belong the wild ass, the zebra, the quagga, the wild and domesticated horse. All animals with this form of foot are most commonly found on level plains, or on the margin of great sandy deserts, for which their feet are the best adapted.

ASSES OF ANCIENT TIMES.—In the Scriptures mention is made of the ass at a much earlier date than of the horse. A member of this despised race was the only quadruped ever honoured with the gift of speech, which was given to him, although only momentarily, to rebuke Balaam, his master, for cruelty. In ancient times, in Oriental countries, a man's riches were often estimated by the number of asses he possessed. We read in the Book of Job, xlii. 12, 'So the Lord blessed the latter end of Job more than his beginning : for he had fourteen thousand sheep, and six thousand camels, and a thousand yoke of oxen, and a thousand she asses.'

It is also stated in the Book of Genesis, that when Jacob sent his sons into Egypt to buy corn they took their *asses* with them. In the Book of Samuel we are told that Kish, the father of Saul, possessed numbers of these animals. In the Book of Judges, v. 10, Deborah, in her song, says, 'Speak, ye that ride on white asses,' etc. In the same book, x. 3, 4, 'Jair, a Gileadite, judged Israel twenty and two years. And he had thirty sons that rode on thirty ass colts,' etc.

In the First Book of Samuel, xxv., Abigail is represented as having laden a number of *asses* with presents for David, and herself as riding upon an *ass* to meet him. The prophet Isaiah, in rebuking the Israelites for their ignorance and in-difference, says, 'The ox knoweth his owner, and the *ass* his master's crib : but Israel doth not know, my people doth not consider.' The prophet Zechariah, ix. 9, under divine inspi-ration, portrays the promised Redeemer of the world as riding into Jerusalem 'upon an ass, and upon a colt the foal of an ass.' According to Luke xix. 35-40 this prophecy was literally fulfilled. It was then that

> A member of this poor and long time-honoured race
> Did bear upon his back the Lord of life and grace.

'This was not necessarily an act of humility, but a peaceful and triumphal entry.' The above quotations prove that the ass in ancient times was by no means considered to be a. despicable animal.

PERSIAN ASSES AND THE SCYTHIANS.—The ass, for various reasons, has been differently estimated, valued, and treated by different nations in all ages. Baird says : 'The ancient Egyptians held the ass in great horror, but the moderns make much use of them, take great care of them, and rear them fine animals. The Indians looked upon them as unclean animals, and held in great contempt those who used them ; whilst the Persians and Arabians made much use of them, as did also the Hebrews. The ass is capable of attachment to his master; has good eyes, a quick scent, good hearing, and is very surefooted. Its cry is peculiar, very prolonged, and discordant. In the expedition of King Darius against the Scythians, it is recorded that the cry of the Persian asses belonging to the king's army frightened the cavalry of the Scythians (among whom that animal was unknown), and made them recoil from the charge.'

'Asses are rare in Sweden and the extreme north of Europe, and though they were introduced to the United States by Washington, they are unknown in the Southern States of the Union.'

Asses and Grand Festivals.—Captain Brown says : 'The ass was anciently unknown in the countries of northern Europe. In Greece and Rome, however, it was held in much estimation, and honoured in their mythology and festivals. By its braying it was said to have discomfited, severally, the deities who warred against the liberty of Jupiter and the chastity of Vesta, and the Ides of June were celebrated in Rome as the festival of the ass. On that occasion banquets were set forth at the doors of the citizens, the millstones were decked with garlands, the asses, which on work-days turned them, were led in holiday triumph, covered with wreaths of flowers, and the grateful ladies of Rome walked before them in the procession, barefoot, to the temple of the goddess whose honour the braying of the ass had saved. The Church of Rome, many of whose festivals were an accommodation of Pagan rites to a supposed subservience to Christianity, formed of the festival of Vesta the Feast of Asses, which, during the dark ages, was held with particular hilarity in Britain.'

On the origin of the domestic ass, *Asinus vulgaris*, authors are not agreed; while some are of opinion that it is a descendant of the wild ass of Asia, others assert that it is doubtful if this species has ever been found in a truly wild state, the wild asses being distinct species.

WILD RELATIONS OF THE ASS.—The following characteristics of the different species of wild asses and the domestic ass are worthy of notice, as tending to show the physical differences existing between them.

The domestic donkey is generally of a grey colour, and has always a longitudinal dorsal streak of a darker hue, with another across the shoulders ; and the ears are long and acute. In wild asses the streak is not so prominent, and the ears are shorter and rounder. A species of wild ass found in Thibet, and known as 'kiang,' usually run in troops of about a dozen, more or less, and generally allow themselves to be controlled in their movements by a male. They are exceedingly fleet of foot, and it is said they can outstrip the Arabian steed. Animals of this species *neigh* like a horse ; the domestic ass *brays*, a noise produced by two small and peculiar cavities at the bottom of the windpipe, which are not found in the windpipe of the 'kiang.'

'The *peechi* (*Asinus Burchellii*) is a fine animal, inhabiting the plains beyond the Orange River, and, in the ears and tail, resembles the horse. It is finely marked, possessing much of the graceful symmetry of the horse, and combines comeliness of figure with solidity of form. Its voice is a shrill abrupt neigh, and has no analogy to the braying of the ass. The senses of smell, sight, and hearing are extremely delicate. When these animals are menaced by an attack from either man or beast, they combine in a compact body, and with their heads placed together in a close circular band, they present their heels to the enemy, and deal out kicks in equal force and abundance.'

ASSES NOT DEGENERATED HORSES.—Nearly as the horse may be considered to be related to the ass, there are in them physical differences. The tail of the horse is covered with long hair to the base ; that of the ass has long hair at the *end* of the tail only. The mane of the horse is usually very long, hanging over at one side of the neck ; that of the ass is short and erect. The horse has all four legs furnished with *warts*, or *sallenders*, while they are found only on the fore-legs of the ass.

It has been asserted that, of all animals covered with hair, the ass is the least subject to vermin ; the reason being, according to an old writer, that one of his ancestors carried the

15

Saviour on his back. The skin is hard and elastic, and is in request for a variety of purposes. The milk of the ass is considered to be very nutritious, and is much used by invalids. It is more easily digested than cow's milk, because it contains less butter and more saccharine matter.

THE USES OF ASSES.—That the domestic ass is a useful animal needs no argument to prove. Many a poor man, both in town and country, would be much poorer than he is if it were not for his donkey. These animals live upon very inexpensive food, and when turned into the fields and lanes, will feed upon thistles and like produce, which the horse and other animals reject. They are hardy and enduring, and, though slow in their movements, are patient in toil. A French writer says that if one horse does twice as much work as one ass, he eats four times more, so that the economical importance of the two animals is not so different as many suppose.

Much as has been said about the stupidity of the ass, he is not devoid of merit, nor entirely without interesting traits of character, which we shall now endeavour to point out, as well as to show that much of the obloquy which ignorant and cruel men have heaped upon him, has been undeserved.

YOUNG ASSES.—There is a charm about the young life of all animals. The babyhood of the donkey is a particularly interesting period of its existence. The animal is then a lively, pleasant fellow, full of fun and mischief. The sight of these creatures brings back to memory the frolics which, in early life, many a boy has had with members of this race in the green lanes of his native home; with what hilarity, after school duties were over, Ned and Tom have mounted the back of the sage old mother, and Frank the back of the infant donkey, for the luxury of a ride without payment, or any thought of the law of humanity which renders overloading an act of cruelty, and amenable to law; and how these animals have, with their noses between their fore-legs, and their hind ones thrown up, landed the would-be equestrians on the hard rough road, and then galloped off, leaving them to bemoan the shaking, the stunning, the bruises and cuts they have deservedly received for their temerity in invading the quietude, the rights, and privileges of donkeydom.

The lives and lot of these animals differ very widely. At one time we may see them the obedient servants of costermongers, at another the willing slaves of country hawkers; now, the companions of wandering gipsies, and then, it may be, in costly panniers and trappings, carrying the children of the high-born and wealthy. But, however and wherever seen, they always appear to be worthy of our notice and admiration.

NORFOLK DONKEYS.—Our finest donkeys are those on the coast of Norfolk, where they are more humanely cared for than is the usual lot of these animals. It is said that the ancestors of the Norfolk donkeys were captured from the Spanish Armada. Be this as it may, kind treatment improves the donkey race in every particular.

It is much to be regretted that, notwithstanding all that has been said and written in favour of the domestic ass, there is no animal more subjected to ill-treatment than he is. His apparent dulness, slowness of movement, his hardy nature, and thick ragged coat, seem to furnish to some men a reason for treating him unkindly, not only in neglecting to groom him and to give him a proper supply of food, but in belabouring him with stout cudgels heavy enough to break his bones.

A man or boy will be willing to do his best for a good master, but will be indifferent to the interests of a bad one. Ill-use a donkey, and he may become vicious; treat him kindly, his intellect will expand, all the good qualities of his nature will be brought out, and he will become one of the most sagacious of animals.

DONKEY BAROMETERS.—Strange as it may appear, it is nevertheless true that donkeys may be regarded as good, living, walking barometers. They seem to be in some peculiar way affected by changes in the atmosphere. They shake their ears and bray before rain, and are particularly disturbed in showery weather. There is no doubt that the electric state of the air before wet produces in some animals a peculiar sensation, which makes the peacock squall, the pintado call 'Come back,' and the ass to bray. An old adage says :

'When that the ass begins to bray,
Be sure we shall have rain that day.'

It may be well, therefore, in hay-making and harvest-time to observe the proverb:

' Be sure to cock your hay and corn
When the old donkey blows his horn.'

15—2

DONKEYS AND SUPERSTITION.—As with many other animals, so with the donkey, ignorance has associated some foolish and superstitious notions entertained until very recently, if they are not so even now, in remote places both in our own and other countries.

Referring to the origin of the streak down the donkey's back and shoulders, Brown, in his 'Vulgar Errors,' says, that from whatever cause it may have arisen, it is certain that the hairs taken from the animal so marked are held in high estimation as a cure for hooping-cough. In the metropolis, as lately as 1842, an elderly lady advised a friend who had a child dangerously ill of that complaint, to procure three such hairs and hang them round the neck of the sufferer in a muslin bag. It was added, that the animal from whom the hairs are taken for this purpose is never worth anything afterwards, and consequently great difficulty would be experienced in procuring them ; and further, that it was essential to the charm that the sex of the animal from whom the hairs were to be procured, should be contrary to that of the person to be cured.

A Monmouthshire paper, some time since, gave the following instance of superstitious belief : 'A patient ass stood near a house, under whose body and over his back a father passed his little son a certain number of times with as much solemnity as if he had been performing a sacred duty. This done, the father took a piece of bread cut from an untasted loaf which he offered the animal to bite at ; nothing loath, the donkey laid hold of the bread with his teeth, and instantly the father severed the outer portion of the slice from that in the donkey's mouth. He next clipt some hairs from the neck of the animal, cut them into minute particles, mixed them with the bread, which he gave to the boy to eat. The donkey was then removed. One of the bystanders was, however, curious to know what the ceremony meant. The father stared at him, and then in a half-contemptuous, condescending tone informed him, that "It was to cure his poor son's hooping-cough, to be sure."'

Quiet and insensible as these animals are considered to be, they have been known to avenge insults by the infliction of serious injuries upon those who have tormented them.

THE ASS AND DRUNKARD.—Some time ago a raving drunkard attacked a poor ass in the stable through sheer malignity, and kicked her so furiously in the stomach that she turned upon him, and bit off his entire upper lip, and left upon the face of this wretch a ghastly memento for life.

THE ASS AND BULL-DOG.—We have read of an ass attacked by a bull-dog seizing his adversary with his teeth in a part of the dog's body which prevented him from retaliating; the ass deliberately carried him to the river Derwent, plunged him deep into the water, and then lay down upon the dog until life was quite extinct.

A PERFORMING ASS.—We know a poor man in Westminster who possesses a donkey of remarkable intelligence. This animal in his younger days (for he is now twenty-six years old) was employed in two of the London theatres to take his part in certain plays which required, and were rendered all the more interesting by, the presence of 'a real live donkey.' He knew the time for appearing before the audience, the precise position he had to take, the particular movements he had to make in advancing, receding, and when to leave the stage. This donkey was not only intelligent but reflective, and must have had a capital memory. His owner told us that he had made twenty-four shillings per week by letting out his donkey in this way, besides what he earned during the daytime.

A GIPSY'S ASS.—One of the best fed, lively, and intelligent members of the ass family we ever saw is one belonging to a gipsy scissors-grinder, who resides in Somerstown. This man is also a mender of kettles and saucepans, and does an extensive trade in the north of London. He goes a certain round each day in the week, which the donkey knows as well as his master. When they leave home in the morning, 'Old Jack' will, without being guided or driven, always take the right direction. 'Jack' seems to have a very commendable idea of the value of time, as he is in the habit of giving his master a hint of the flight of it when he stops to gossip with anyone in his rounds. This hint consists in a side-long look and a subdued noise, half resembling the grunt of a pig. If the gipsy doesn't take this hint, then 'Jack' begins to shake, not only his harness, but the whole of the kettle-mending and scissors-grinding paraphernalia, and to paw with his fore-feet, as much as to say, 'If you don't come I shall be off.' A gentle word from the humane gipsy soothes 'Old Jack's' perturbed spirit, and he jogs on, evidently much more satisfied in doing so than in waiting for idle gossip and doing no business.

A Wonderful Ass.—There are but few animals whose powers of endurance are greater than those of the ass, of which we have an example in the following anecdote: 'In 1826 a clothier of Ipswich undertook to drive his ass in a light gig to London and back, a distance of 140 miles, in two days. He did so, having travelled at the rate of seven miles an hour. The animal was in no way distressed, and he performed the journey without the use of the whip. He was twelve and a half hands high, and half-bred Spanish and English.'

A Sensible Ass.—In a by-lane, not far from a large town in the West of England, two dogs had a terrible fight, in which they were urged on by some fellows of depraved habits and brutal instincts. One of the dogs seized the other and retained his grip so firmly that the poor victim could not shake him off; a donkey, browsing hard by, hearing the cry of the dog, pricked up his ears, rushed forward, and, breaking the circle of biped spectators, seized with his teeth the tail of the other dog, held him up, and then gave him such a shaking that he was compelled to relinquish his hold. The ass then resumed his feeding. The dogs tried to escape, but were again set on to fight. That the ass showed better sense than did these so-called 'lords of creation' is beyond all dispute.

Donkey Show.—A most interesting exhibition of donkeys took place in the Crystal Palace in May, 1875, respecting which a London newspaper said: 'Though a strong man could pick up any one of them and carry it off without any very great exertion, most of them are credited with being able 'to draw one ton.' Another phase of the donkey's character is given in the *Animal World*, which refers to the same exhibition: 'Some of the animals looked extremely mischievous when approached by strangers; but upon the arrival of the owner, reciprocates were exchanged which argued well for the genera treatment and care bestowed upon them. Even if any member of the owner's family appeared on the scene, the donkey, upon recognising him, appeared quite sure that everything was right. In one stall a donkey had been apparently frightened, and its excitement was causing some amusement to the mischievous, as well as terror to the timid, when two little children ran up to the animal. Everyone anticipated broken bones; but to our great surprise the excited animal immediately showed signs of recognition, and permitted the children to fondle it, its fright being immediately transformed into clumsy, but tender, playfulness. The owner then came up to the stall, and upon his attention being called to what appeared the dangerous proximity of the children to the donkey, he replied, "Ah, he knows the children as well as he does me; he won't hurt 'em!"'

CHARACTER OF THE ASS.—The following eloquent description of the character of the ass is given by Buffon: 'He is naturally as humble, patient, and quiet as the horse is proud, ardent, and impetuous; he suffers with constancy, and perhaps with courage, chastisement and blows; he is moderate both in the quantity and quality of his food; he is contented with the hardest and most disagreeable herbs, which the horse and other animals will leave with disdain. The ass has, like all other animals, his family, his species, and his rank; his blood is pure, and, although his nobility is less illustrious, it is equally good—equally ancient with that of the horse.'

In affection, not only for its master but for its own progeny, the ass is not wanting. Pliny says: 'That when the young is separated from the mother she will pass through flames to rejoin it.' M. Bourguin says: 'The donkey is a model of sobriety, meekness, and resignation; he asks for nothing, abuses no privileges, but lightens the heavy burdens of weary life by his own patience.'

If, as we have stated, the ass is so prominently mentioned in the Scriptures—if his structure is so wonderful, and his uses so great—if he has so many good qualities, and strong affection, as well as being man's cheap and willing slave,

> 'Why should this creature be ill-housed, ill-fed,
> That for so many thousand homes wins bread?
> Treat the ass well, and merrily he'll go;
> A *gentle pat* is better than a blow.'

A Hunter and a Racer.

CHAPTER XIX.

EVERYBODY'S FRIEND.

Thy graceful form, thy strength and speed,
Thy fine dark eyes, and flowing mane,
Thy uses too, tell man thou art
A noble prize for him to gain.

HERE are but few things accomplished by man in which he has shown his intellectual power more fully than in his success in subjugating animals of almost every kind, and making them subservient to his own purposes.

The most savage carnivora—powerful elephants, huge amphibia, and many other forms of life whose native homes are the wild woods and rolling rivers of Asia and Africa—are, as we see in our Zoological Gardens, imprisoned, and there remain under man's control and at his pleasure. Even the mighty leviathan of the deep seas is captured and utilized by him in various ways.

But there is no animal conquered by man, which, taken altogether, has ever been rendered so useful to him as the horse, who is directly or indirectly everybody's friend, as there is but little business transacted or pleasure taken that is not in some degree identified with this noble animal, whose structure, habits, and intelligence are alike very wonderful.

THE HORSE.—The reader may ask, 'What more can be said about the horse than has already been written? Is not man's acquaintance with him of very ancient date, and is he not common in almost every country in the world, and in the possession of the richest, the poorest, the highest, the lowest, the most refined, and untutored of the human race ; and is he not closely identified with the life, business, and pleasures of man everywhere, and at all times ?

All this we admit, and would observe that we do not aspire to any *new discoveries,* either as to the structure, instincts, habits, or uses of this animal. Our object is to give, in as concise and popular a form as possible, some information respecting him, with the hope that a feeling of interest may be awakened on his behalf, and that we may be induced to place a proper value upon him as one of the greatest friends of man.

While it is true that many animals excel the horse in size, strength, and herculean proportions, he stands unrivalled for symmetrical form and beauty, majestic appearance, high-spiritedness, fleetness, nobility of character, general utility, and as combining in his organism some of the greatest marvels of animal life, which we now propose to describe.

THE HORSE AND HIS ORIGIN.—The *Horse* belongs to the class *Mammalia,* order *Ungulata,* family *Equidæ.* The genus *Equus* is the type of the family, and till lately constituted the only genus belonging to it.

There is some difficulty in determining the native country of this useful animal. Some writers think we are indebted to Arabia for the horse, but it has been observed that our earliest accounts of the horse are from Egypt, and that we have no mention of it by historians as being in Arabia till after the time of Mahomet the prophet.

Be this as it may, it is well known that no people on the face of the earth treat their horses with more care, kindness, and affection than do the Arabians. Two reasons may be assigned for this ; the first is a recognition of the great utility of the horse to them in their peculiar circumstances, and predilections in favour of a wandering mode of life ; and the second, and no doubt the stronger reason, may be the notion they entertain of the origin of their much-loved animals.

The following information bearing upon the last reason referred to, with which we have been favoured on good authority, may be read with interest.

General Danmas having addressed a letter to Abdel Kader to know his opinion of the origin of the Arab horse, received the following statements in reply :

' God created the horse out of the winds, as he created man

out of the dust. Many prophets have proclaimed that when God would create a horse He said to the south wind, " I will bring out of thee a creature, be thou therefore condensed." Then the angel Gabriel taking a handful of the matter presented it to God, who formed therewith a brown-bay horse and said, " I name thee *Horse*, and create thee Arab, and give thee a bay colour. I attach blessing on the forelock ; thou shalt be lord of all animals. Thou shalt fly without wings, and from thy back shall proceed riches." Then marked He him with a star in the forehead, the sign of glory and blessing.'

AFFECTION OF HORSES.—That love and affection beget the same feelings in those on whom they are bestowed is generally as true as that light comes from the sun. It is said that a wandering Ishmaelite, who treated his horse as if he had been a child, found, when he was a captive, that his faithful steed bore him by his teeth to a place of safety and freedom, and then died a willing martyr at his feet. With such an exhibition of generosity, devotedness, nobility and faithfulness on the part of the horse, we need not wonder that the Arabians should love this animal.

A celebrated historian informs us that ' Among the wandering tribes of the predatory nations of antiquity, the services of the horse were indispensable. These lived in the open air, subsisting on the coarsest food, performing long journeys through uncultivated or hostile countries, generally on horseback ; their wives and children followed in waggons dragged by horses. They seldom dismounted, but eat and slept on horseback. The horse was still further serviceable to the barbarous Samaritans ; they ate its flesh and drank its blood mixed with the milk of sheep. Yet these horses were carefully reared, of an excellent breed, and, as Pliny says, capable of performing a journey of one hundred and fifty miles on a stretch.'

PARTHIANS AND THEIR HORSES.—' Of all the nations of antiquity the Parthians are the most commonly celebrated for their superior skill in the management of the horse. They cultivated with great attention the breed which was noted for the lightness of the colour of the eyes, and for having the one eye generally differing from the other. The horse was trained to obey the slightest motion of the rein, and to change with the utmost rapidity from one direction to another. . . . .

As the Parthians employed the horse in war, so the licentious Sybarites associated it with their pleasures; they taught their horses to dance to the sound of pipes, and introduced them as an amusement at their common feasts.' It appears that the Crotonians in a war with the Sybarites sounded the strains to which the horses had been accustomed, which caused them to dance, threw them into confusion, and brought about the defeat of the Sybarite army.

We are informed that the true birthplace of the horse is the high central plateau of Asia, in the north-east chain of the Caucasus. It is very probable that those herds of wild horses now found in America are the descendants of some horses that were taken to that country from Europe about 1537, which, being deserted, ran wild, and lived in troops, which were usually guided by a male. They now exist in large numbers, and occupy an immense tract of country extending from Buenos Ayres to the Strait of Magalhaens.

WILD HORSES.—In an interesting work on the quadrupeds of Paraguay, D'Azara, referring to wild horses, says they are very numerous in South America; he assures us that it is not uncommon to meet with troops of thousands of them. They are both troublesome and destructive, because they not only consume the best of the pasture, but attract the tame horses that approach by their neighing and caresses; and when once a horse has joined them, he is lost for ever. On this account, travellers are obliged to be on their guard, that they may not lose their horses; and, therefore, whenever they perceive a troop of these wild animals, they attempt by every means to frighten and force them away.

When they advance, it is not in line of battle, but one of the foremost is despatched, and the rest follow, forming a column without any interval, and which nothing can break. Sometimes they make several circuits round those who attempt to frighten them, and sometimes only one, before they set off and disappear.

Baird informs us that 'On the Pampas, the Guachos, a semi-civilized race of men, live amidst these horses, and their method of capturing and breaking them in is very curious. It is said they can secure and break in one of these young horses in the course of an hour.'

According to Herodotus white savage horses were, in his

time, to be found in Scythia; and he adds that beyond the Danube there were wild horses covered with hair five inches long. Other ancient writers speak of wild horses in Syria, Spain, and the Alps; on the island of Cyprus, the Cape de Verde, as well as in the deserts of Africa and Arabia.

WILD HORSES AND BIRDS OF PREY.—Half a century ago birds of prey were employed to hunt wild horses in the country bordering on the Caspian Sea. These birds would alight on the horses' necks, and by clinging to them and tormenting them compel the animals to run until they were exhausted, and so making the capture of them an easy matter. Wild horses are defective of that symmetry and beauty which characterize many breeds of the domestic horse. The former have long shaggy hair, are somewhat ill-formed, have large heads, and are by no means well-proportioned animals.

Of domestic horses there are many kinds, differing materially in size, strength, and colour. In cold regions they are small and rough-haired; but in southern climates large and sleek. We have also many varieties of them produced by the mixture of breeds.

THE DRAY OR DRAUGHT HORSE.—These horses are usually of enormous size and great strength, and are employed by large brewers, distillers, coal merchants, railway companies, and carriers. Some of them are capable of drawing, with comparative ease, on a good, level road, two tons each. They are the produce of the old English draught-horse mixed with the Flanders breed.

THE COACH HORSE.—These horses derive their origin from the hunter and a common mare. Some of them stand very high, are of great muscular power, and tolerably fleet of foot. When well fed they are very high-spirited, but gentle when properly treated.

THE HACK.—This horse is used for the saddle and common road work. The Suffolk sorrels, and the bays from Durham and Yorkshire, are esteemed the best. The pack-horses in the latter county were at one time extensively used in conveying manufactures to remote parts of the kingdom, often over the highest hills of the north, as well as on level roads. Their use, however, in this particular, has been superseded by our railways and other more recent means which have been adopted for the transit of both passengers and goods.

THE HUNTER.—This horse is related on the one side to the pure blood racer, and on the other to the female of the last-mentioned hack, or roadster. Hunters are much fleeter than the hack, and much stronger and able to bear more fatigue than the racer. The powerful muscles and the elastic tendons of this animal are in all respects not only required but admirably adapted for the exciting chase over hills and dales, hedges and ditches.

THE RACEHORSE.—This is the highest bred of the horse family, and may boast of the purest blood, at least of those in our own country. The racehorse owes much of its fleetness and endurance to a mixture of the Arabian blood, which is supposed to have first taken place some time after the Crusades. Many racehorses in England have distinguished themselves by winning immense sums of money, plate, and other things, sometimes to the value of £10,000, and even 15,000 guineas.

PONIES.—These are horses of a smaller breed than the before-mentioned ones. Some of them are scarcely larger than a Newfoundland dog. The most diminutive is the Shetland pony. There are also the Forest of Dean and Welsh ponies, rather larger than the Shetland breed, and much used for drawing very light traps and carrying children. Great numbers of these ponies are sold by gipsies at the fairs in the western counties of England. They are quick, half wild, and usually have long matted hair. They, however, become, by proper care and kind treatment, very gentle, docile, and handsome creatures.

Horses, of every breed, are the most useful animals man can possess. In life they serve every purpose, both in labour, profit, and pleasure; and even after death their hoofs, hair, and skin are articles of commerce.

HORSES IN BRITAIN.—About a dozen years since there were in Great Britain about 2,000,000 of horses. If we reckon these to be worth, on an average, £20 each, then we have in horseflesh the value of £40,000,000 sterling. If these horses bring in only three shillings per day each, then, during 300 days in the year, they would earn £90,000,000. Of course a considerable amount of money must be allowed for expenses, say ten shillings per week for each horse (£52,000,000); there would then be, by the work of these animals, a clear profit of nearly £40,000,000. But if, instead of earning only three shillings per day each, they were to bring in six shillings per day, the clear profit would be nearly £80,000,000.

If each of this number of sovereigns could be beaten out to cover one square foot, then all of them, so beaten out, would spread over 80,000,000 square feet, which are equal to nearly 1,837 acres in extent. Cut this sheet of gold into a ribbon one inch wide, it would be long enough to go four times round the world at the equator, and three times round it north and south, and 386 miles of ribbon left to make a rosette for John Bull's best button-hole.

The above calculations apply to our own horses only. To say nothing of the millions of horses in continental countries, how immense is the value of these animals, and how much we owe them in various ways! Some of our successful race-horses have been sold at almost fabulous prices, even as high as £10,000, and sometimes more.

Should the above illustrations and curious calculations be considered unnecessary and useless, so far as a description of the horse is concerned, they show at least that the great numbers of this animal prove their necessity, and that this necessity cannot be dispensed with in our present state of civilization.

From a valuable work on 'The Horse,' published by the Royal Society for the Prevention of Cruelty to Animals, we gather the following interesting and useful information on the foot, eye, ear, and stomach of the horse :

'THE FOOT.—To most persons the foot of the horse appears to be only a roundish hard lump of horn, on which an iron shoe is nailed to prevent its being worn away by the roads. Such persons may perhaps hear with astonishment that it is a complete and elaborate instrument, perfectly adapted to the work it is intended to perform. . . . The real foot of the horse is enclosed in a horny case called the *hoof*, the outside rim of this casing forms what is called the crust, or wall. . . . In the hinder part of the foot is an elastic substance called the *frog*, which forms a soft and yielding cushion on which the horse's foot partly rests, being thus relieved from the shock of the hard hoof on the ground. . . .

'If the foot were an unyielding mass, the danger of slipping would be great. But, instead of this, it has a prominent edge all round, which takes a firm hold of the ground and obviates the difficulty. This hoof is elastic, and, on the weight of the horse being fully thrown on it, allows the inner soft cushion, or frog, to descend and press firmly and tightly on the earth. Thus two ends are attained—firmness in the tread, ensuring the horse's safety, and a regularity of pressure which obviates the jarring that would be so painful and prejudicial.

'When the animal is in a state of nature, its hoof is strong enough to need no artificial protection, but on the hard and stony roads common in all civilized countries, it has been found necessary to fit something to the foot to protect it from the great wear and tear which it unavoidably incurred. For this purpose nothing has been found so effectual as what is termed *shoeing*, or affixing a thin plate of iron round the outer hard and horny edge of the hoof—a practice known in Britain during the time of the Romans. When done with judgment, the proper action of the foot goes on nearly as usual ; but, if injudiciously performed, the action of the horse is impeded, lameness is caused, and temporary or permanent diseases are brought on.'

Although we do not presume to point out to the farrier the rules to be observed in horse-shoeing, yet taking into consideration the delicate organism of the horse's foot, it is of vital importance to the comfort and utility of the animal, as well as to the interest of his owner, that special care should be taken in the matter of shoeing. Many a valuable horse has been ruined by men, who, through ignorance of the structure and requirements of the horse's foot, have by cutting, paring, and unnecessary rasping, destroyed the provision nature has made for the use and comfort of the animal, often rendered him comparatively useless, lessened his value, and inflicted upon him unnecessary suffering.

If a connoisseur in painting were to ask his man-servant to recommend to him a person to clean and restore an old family picture worth a hundred pounds, and the man were to tell him of a common signboard painter, the master would no doubt ask him if he thought he was mad, or a lunatic, to allow so valuable a picture to go into the hands of the person he had recommended ? And yet it is by no means unlikely that the same gentleman would be less thoughtful in the matter of horse-shoeing, and might, without one word of inquiry as to the competency or qualification of the farrier, commit to his care, to be shod, a horse worth three or four times more than the picture referred to ; and if asked his reason for so doing, would simply say, because the farrier selected lived near, and it was more convenient to send his horse there than a greater distance off. When it is remembered that the utility of the horse depends entirely upon the health, good order, and soundness of his feet, the question of good shoeing becomes of paramount importance.

THE EYE.—On the eye of the horse, the work before referred to says : ' The eye of the horse differs in some points from the eye of man, and it has some appendages not possessed by the latter, which are required by the configuration and habits of the animal.

'The horse has no eyebrows, and his eyelashes are arranged in a peculiar manner, the longest hairs being on the upper lid, probably that the eye may be defended from excess of light and from insects, which would naturally endeavour to annoy the horse in that unprotected part. . . .

'On the lower lid of the eye are some long projecting hairs or bristles, which are supposed to be useless by ignorant persons, and are sometimes cut away. Are they, however, useless ? Far from it ! They are intended to let the animal know the presence of anything that may approach the eye too closely. If the reader will touch one of these hairs, and observe the sudden twitch and closing of the eye, he will be able to appreciate the importance of these supposed useless and superfluous excrescences.

'The horse has no hands wherewith to rub his eyes when they are irritated by dust or similar substances. A continual drying of the liquids which moisten that part is constantly going on, more especially when moving quickly along ; and the Almighty has, therefore, in His wisdom, provided an efficient substitute for so necessary a purpose. Just inside the upper lid is a little organ called the *lachrymal gland*, which is continually sending out a liquid to flow over the eye, and wash away all lesser impurities. Besides this provision there is a thin cartilage or membrane concealed in one corner of the eye, vulgarly called the *haw*, and this, whenever the animal wishes, can be pushed out along the surface of the eyeball. The dust, or the insect that may be the cause of the irritation, wet with the tears, is swept up by this membrane, and immediately carried away.'

The EARS of the horse are easily turned to the front or the back of him, and he has the facility of reversing them by pointing one of them forwards and the other backwards at the same moment. His eyes are placed at the *sides* of his head, so that he can easily perceive objects approaching him even from behind. His upper and lower lips are furnished with long hairs, which serve as whisks to brush away flies and other insects which are apt to tease him while feeding.

The STOMACH of the horse is very small in proportion to his size, and much less in proportion than man's ; he cannot, therefore, take much food at once, but requires often feeding. The *lungs* and *stomach* of this animal are separated by a partition of thin wide muscle, and both lungs are much larger when in use. As the lungs occupy much more room when the horse is moving about, the space they occupy is then so filled that only one of them can be distended at a time. The horse can distend his *lungs* and breathe hard, trot, or gallop fast, provided his stomach be empty ; he can fill the latter with safety when at rest, or nearly so, till the food is digested. But if they (the lungs and stomach) are both full the greatest danger is to be apprehended ; the horse is sure to be ' blown ' almost immediately, because he has no room to breathe, and rupture of the stomach may cause the animal to drop dead in a minute.

The horse possesses no gall-bladder, because the process of digestion is almost incessant, and the bile passes off as rapidly as it is formed.

USEFUL HINTS TO THOSE WHO HAVE THE CARE OF HORSES.

1. Make yourselves acquainted with the structure, physical powers, utility, intelligence, and moral characteristics of your horses.

2. In feeding your horses take care that the food given to them is sweet, and varied according to their tastes and powers of mastication ; but do not run or work them hard immediately after feeding, or apoplexy may follow. Remember the rhyme :

> ' Full feed, then rest ;
> Often feed does best.'

Four half-pails of water per day are considered to be sufficient for one horse.

3. Let your stables be dry, well ventilated and light, because a sudden transition from a dark stable to a strong light sometimes affects the eyesight very seriously.

4. Do not keep the fodder in a loft over the stable, because the effluvia arising from the latter place, and passing upwards to the loft, may so impregnate the food as to render it unwholesome, if not injurious.

.5. Do not place the hay-rack so high as to compel your horses to reach up at neck-length to pull the hay down, because by doing so dust may get into their eyes and cause inflammation and great pain, and because this system of feeding is also productive of roaring. Horses naturally feed upon the ground, therefore so place the hay that your horses may pick it up, not pull it down.

6. Never allow hen-roosts to be in, or near, your stables, because fleas, by which fowls are usually infested, are apt to lodge also in the hair of horses, to cause great irritation in their skins, and, as a natural consequence, to make the animals so uneasy that they can get neither rest nor sleep, which unfits them for work.

7. If, when on a journey with a load of any kind, you stop at some wayside inn or other place to refresh yourselves, attend to the needs of your horses as well as to your own; and while they are standing waiting for you let the rest-stick be down to support the shafts of your cart, so that the backs of your animals may be relieved from the weight of the load behind them. Take care also that the wheels of your cart or waggon be well greased; this will lessen the labour of your horses, and help them to do their work with greater ease. Never, on any account, overload your horses.

8. Avoid that bad habit, in which many young drivers indulge, of giving their horses a cut with the whip when they are walking or trotting along at a proper pace, and then the moment they answer to it pull them violently up by the reins, thereby often hurting their mouths, needlessly exciting and worrying them, as well as helping to spoil their tempers. How can horses be expected to understand what they are to do, or to be obedient, if treated in such a cruel and insane manner? They can be taught to obey a word as well as to answer to the whip.

9. When you wash the legs of your horses rub them until

they are dry, or grease and sore heels may be the consequence. Do not put tar on their hoofs, because it will stop up the pores and prevent humours, generated in the foot, from escaping, and then inflammation may take place and lameness ensue.

10. Do not groom your horses in the stalls where they feed, because the dust, etc., from their bodies may foul their cribs and food, which horses have been known to refuse and to turn away from on that account.

11. See that the harness of your horses fits properly, especially the collars and saddles ; and when these get wet see they are well dried before they are put on again, because when put on wet they are apt to break the skin and to produce sores. Take care that the surface of the collars and saddles is smooth; lumps in either give great pain to the wearers.

12. Never allow the farrier, when shoeing your horses, to cut away more of the frog and the bars than the ragged projecting pieces, because the frog is the elastic cushion for their bodies to rest upon, and nature intended it for the comfort of your horses. Although iron shoes are indispensable to working-horses, the calkins of them should not project more than half an inch, because the higher they are the more they interfere with the proper action of the tendons, bones, and joints of the feet and legs, and may, by throwing them into an unnatural position, bring on weakness and lameness.

13. Never flog your horses for shying ; they often do so because, seeing some object but imperfectly, they do not understand it, and therefore get nervous and frightened. To the use of blinkers many cases of shying are to be attributed.

14. Never, if you can avoid it, attach to your cart a fore-horse lower in stature than the thill, or shaft horse, because, when the former pulls, he brings down a greater weight on the back of the latter, whose labour is made all the greater by this foolish plan. If you must have a little horse in the front, then let his traces be long enough to hook on to the hind part of the shafts, which will throw them more in a horizontal line with the fore-horse's collar, and so avoid the evil referred to.

15. Do not bawl at your horses, it has a bad effect on them ; —speak gently, it pays best ; and remember that the pleasure and profit to be derived from them will be just in proportion as they are properly treated by you. Horses well know who are their friends.

CHAPTER XX.

ANECDOTES OF EVERYBODY'S FRIEND.

Old 'Sorrel' and 'Blossom,' 'Jet,' 'Smiler,' and 'Dan,'
Were five as good horses as e'er drew a van ;
They all were good tempered, quick, willing, and free ;
In fact better horses there never could be.

HAVING briefly described the wonderful structure of the horse, who is *par excellence* man's best servant, because he contributes so largely, and in so great a variety of ways, to his pleasure, convenience, and profit, we shall now say something respecting his higher nature. Writers on natural history and others differ in opinion as to the mental capacities of the first order of vertebrated animals. While in all cases the horse is admitted to be very intelligent, some think he is less so than elephants and dogs, and even than cats and foxes. Be this as it may, it is well known that the horse has a large brain, a good memory, and both imitative and reasoning powers.

Youatt, speaking of this animal, says : 'The horse, with all his noble faculties, and powers, and inclinations, is perfect in the situation in which he is placed.' We think the following anecdotes show clearly that the mental capacities of the horse are of no mean order, and that we have not given him credit for more than he deserves.

THE CAVALRY HORSE AND THE SOLDIER.—A German cavalry soldier and his horse were captured in the fight at La Bourget, and taken off with other prisoners. Three days after the fight they halted for the night in a village. The poor fellow was sitting near the window thinking how he might escape, while his noisy captors around the fireplace were

The English Cart Horse and Welsh and Shetland Ponies.

lulling themselves with wine ; suddenly he hears in the streets the neighing of a horse. His soul is trembling, and his blood stops for a moment No doubt it is his brave steed that had broken loose from a shed where she had been placed and is in search of her master. One of the panes of the window was replaced by paper. Boring with his finger a hole in it, he lays his mouth to the opening, calling cautiously and coaxingly, ' Lizzie ! Lizzie !' A joyous neigh is the reply, and Lizzie is close to the window. In a moment the whole frame of the casement is smashed, and before the tipplers know what is the matter he is outside and on the bare back of his faithful mare. It seems that the sagacious mare knew that the life of her master was at stake, for she runs off like a whirlwind ; and yet she is not urged on by spurs or bridle, for the captors have taken the boots of the rider, and the bridle is hanging by the saddle in the shed. Shots are fired after them, and bullets fly past their ears without stopping the horse. The hussar does not know the way, but Lizzie remembers it ; and after thirty-five hours both arrive at the outposts of La Bourget, happy to be again among their comrades.

THE HORSE AND THE DRUNKEN FARMER.—Mr. Morris says : ' A farmer who lives in the neighbourhood of Bedford, and regularly attends the markets there, was returning home one evening many years ago, and being somewhat tipsy, rolled off his saddle into the road. His horse stood still ; but after remaining patiently for some time, and not observing any disposition in its rider to get up and proceed farther, he took him by the collar and shook him. This had little or no effect, for the farmer only gave a grumble of dissatisfaction at having his repose disturbed. The animal was not to be put off with any such evasion, and so applied his mouth to one of his master's coat-laps, and after several attempts, by dragging at it, to raise him upon his feet, the coat-lap gave way. Three individuals who witnessed this extraordinary proceeding then went up and assisted him in mounting his horse, putting the one coat-lap into the pocket of the other, when he trotted off and safely reached home.'

DECIDED BY THE HORSE.—A gentleman, known to the writer, and a friend had a journey of seven or eight miles to perform along a high-road to a town which both had, with the same horse, travelled to many years before. They had not proceeded far when it became quite dark, and the difficulty of the journey to them was increased by the recollection that farther on the road divided, one road to the right and the other to the left, and running for some distance collaterally with each other. 'We must take the *left-hand* road,' said the one. 'No,' replied the other; 'we must take the *right-hand* road, or we shall get many miles out of our way.' A warm dispute arose on account of this difference of opinion, and neither of the men felt inclined to give way to the other. That both could not be right was certain. As they met no one of whom they could obtain the needed information to decide the dispute they were in a great dilemma, and knew not how to act. They, however, at last agreed to let the horse take whichever road he pleased. So slackening the reins, and without attempting in any way to control the animal, they went on; the horse took the left-hand road, and in about half an hour the travellers found themselves in the town they wished to reach.

A GOOD MEMORY.—We well remember an old horse named 'Sharper' going a journey of about thirty miles to a watering-place, with a cart-load of household requisites for a family who intended to spend the summer season by the seaside. Old 'Sharper' was driven by a man who, although not a drunkard in the common acceptation of the term, was by no means a member of the temperance society. He had a very high opinion of 'John Barleycorn,' and strongly believed in the virtue of a '*drop o' yal*'—that is, *good home-brewed ale*. The driver and 'Sharper,' however, returned home all right, and neither the worse for the journey. Twelve months afterwards the same family went again to their seaside residence, and old 'Sharper' the second time took to it a similar load to that he had done before, but in the care of another driver. In this journey the animal positively, of his own accord, halted at every inn by the road-side where the first driver had taken his glass or two of ale; and it was not until after some minutes had elapsed that he could be induced to move on. When the first driver was told of this he admitted that he had stopped at all the inns mentioned; that 'Old Sharper' was quite correct, and that he evidently possessed a very good memory.

THE MILKMAN'S HORSE.—The following story is also illustrative of the sagacity and memory of this animal: 'Mr. Jones, who intended taking his wife out for a drive one day, asked his milkman (who had a very spirited horse) for the loan of the same, which request was granted. However, Mr. Jones was not a good driver, and had great difficulty in managing the horse, which at last became ungovernable, and, to the great horror of Mrs. Jones, bolted with them. Mr. Jones did not know what to do, and a serious accident seemed unavoidable, when, all of a sudden, Mr. Jones, remembering the capacity for which the horse was used, called out with a stentorian voice, ' Milk oh ! milk oh !' the horse stopped instantly, to their great joy, at this familiar cry, and Mr. and Mrs. Jones got home safely without any further incident, save that when they returned home in the evening, on passing a pump in the neighbourhood the horse would not stir an inch until Mr. Jones got down and worked the pump-handle a dozen times, after which operation it moved on directly ; and, to finish off the day's pleasure, it stopped at all the customers of the milkman in the road where Mr. Jones lives, his house being at the farther end.'

THE GENEROUS HORSE.—'A horse belonging to M. Frederic Cuvier being only allowed to eat straw, on account of some malady, his companions, feeding at the same manger, pushed some of their hay to him every time they received a supply, and the same horse was one day detected in pulling hay out of his rack in order to feed a goat in the same stable as himself. He was as fond of sugar as many bipeds are, and when his master was pleased with him he would give him some ; often did the cunning animal play the same tricks over again, and stand still between each, in expectation of the accustomed reward, turning his head round to see if it were coming. M. F. Cuvier generally provided himself with some lumps of sugar when he took the horse out ; but if he neglected this, stopped at one of the small inns by the roadside, where he procured the reward for his steed, and never again did the latter pass these houses without asking for more.'

A NOBLE HORSE.—If noble deeds deserve a noble title, then we award it to the animal referred to in the following short story : 'A horse was dragging a load of coal down a narrow lane in the neighbourhood of New York; the driver was behind chatting with a neighbour ; the horse, walking slowly along, came to a child sitting in the middle of the road, and then suddenly stopped; not feeling inclined to injure the child the horse waited, and seemed to cogitate as to what was to be done; his sagacity came to the rescue. As there was no room for him to turn aside, he gathered up the child's frock between his teeth, lifted him gently up, and placed him just at the outside of the wheel-track, and then went on, leaving the child uninjured.'

INTELLIGENCE OF HORSES. — In 'Records of Animal Sagacity,' the author of that work gives the following information: 'A gentleman had a horse which, after being kept shut up in the stable for some time, and turned out into a field where there was a pump well supplied with water, regularly obtained a quantity therefrom by his own dexterity. For this purpose the animal was observed to take the handle into his mouth and work it with the head, in a way exactly similar to that done by the hand of man, until a sufficiency was procured.'

THE HORSE THAT HELPED HIS MASTER OUT OF A DIFFICULTY.—At the time when excitement ran high respecting the Australian gold-fields, a gentleman with whom we are acquainted was induced, partly by curiosity and a wish to improve his health, to visit that country, where he remained several months. Having plenty of money, he could indulge in many luxuries which gold-diggers and other classes of working-folks were not able to obtain. One of these luxuries was a saddle-horse, on which he would make short excursions into the country, always returning home before nightfall.

One charmingly fine day, however, he wandered farther away than usual, being attracted by the beauty of the scenery and various productions of nature which were entirely new to him, and by two or three small groups of natives, whose appearance and habits so arrested his attention that he was quite heedless of the swift flight of time; in fact, so pleasantly had it sped on with him, that when he thought of returning he found it was much later than he expected, for daylight was already declining, and he knew that darkness ere long would come rapidly upon him.

He also knew he was several miles from home, but had not the remotest idea as to the direction he should take to reach it. 'Never,' he said to us, 'shall I forget the feeling of isolation and the fearful forebodings which seized me as I sat on the back of my horse, not knowing which way I should go. Everything at that moment seemed to be to me a blank—dark, dismal and hopeless. Being ignorant of my whereabouts, and knowing there were natives in close proximity who might be no more scrupulous about taking my life than that of a kangaroo, I felt almost driven to desperation and to madness.

'To make matters still worse, I remembered there was really no proper road or track of any kind in those parts over which I had travelled; and as I had not in my outward journey marked any object to assist me on my return, I found myself in a *terrible predicament.*

'Just as I had reached the climax of misery, and was about to give myself up, not only for lost, but as the helpless victim of a stern and cruel fate, like a cheering ray of light came the recollection that horses, possessing marvellous powers of memory and of recognition, have been known to extricate their riders and themselves out of circumstances of extreme peril and difficulty.

'So, throwing the reins on the neck of my horse, and giving him a few gentle pats and kind words, I said at last, " Now then, old fellow, do your best in finding your way home." We started, and, strange to say, my guide never once hesitated, but went on and on, carefully picking his way, being, not long after we began our journey, assisted in so doing by the pale light of the glimmering stars. Two hours of uncertainty and fear had passed away when to my great relief I discovered we were near home, which, in less than another half-hour, we both reached in perfect safety. Not to me, but to my noble, faithful, sagacious horse is due the credit and the praise of my getting out of the difficulties I have just narrated.'

It is almost needless to say that horses who are so useful and intelligent should be kindly treated, and every effort made to remove the causes of their sufferings.

ACTING THE OLD SOLDIER.—A few years since a kind-hearted man living in the north of England had a brood mare, commonly called 'Old Sal,' who was subject to those griping pains with which horses are now and then afflicted. When ' Sal ' had these attacks her owner had recourse to the usual remedies, which always produced the desired effect. On one occasion, however, ' Sal ' and her master were a long distance from home when unfortunately the former was seized with her old complaint, which was more severe than usual, as the mare trembled and seemed to be in great agony. Her humane master feared she would die on the spot; and as he had nothing with him to administer to her as on former occasions, he allowed an eye-witness of these sufferings, who had obtained half a pint of gin at a village inn, to give it to the mare. The effect appears to have been almost marvellous. 'Old Sal' became easy, and in a little time she and her master resumed their journey homewards. On many subsequent occasions, when the mare was attacked as described, gin was given to her, always with the same results as before. It was soon, however, observed that these attacks had become unusually frequent, and that the gin-doctor was getting very expensive. 'It was evident,' said 'Sal's' master, 'that she had acquired a liking for this stimulant, and that she feigned illness to get it when there was little or nothing of a serious nature the matter with her.' In fact, he had so many reasons for entertaining this idea that he often accused 'Sal' of 'acting the old soldier.'

SAGACITY OF A CART HORSE.—A writer in the *Animal World* says: 'Directly opposite my residence a church is being erected, and during its progress temporary sheds have been put up, for the use of the workmen, and one as a stable for a very fine cart-horse, the property of the builder. The extreme docility of this animal attracted my attention to him, and since that some of his manœuvres appear to me to border strongly on the sense and the powers of reflection. His stable was erected at one end of the church; on one occasion two poles had been fastened across his usual road to it, in order to strengthen the scaffolding; he went up, tried the strength of these first, then finding that he could neither get over nor under, he turned round, and, at a full trot, made the circuit of the church, and got to the other side of the poles by another path. Here was no straying about, and at last finding his way, he resolved to go round as if an idea had at once flashed across his mind. Another day, a waggon had been left standing in the narrowest part of his road to the stable; he looked and tried each side, but found there was not space enough for him to pass; he took very little time for consideration, but put his breast against the back part of the waggon, and shoved it on to a wider part of the road, then deliberately passed on one side to his stable. Could human wisdom have done better? But to crown all his manœuvres, I mention the following as being, I consider, very extraordinary: During the winter a large wide drain had been made, and over this strong planks had been placed for our friend, the cart-horse, to pass over to his stable. It had snowed during the night, and froze very hard in the morning. How he passed over the planks on going out to work I know not, but on being turned loose from the cart at breakfast, he came up to them and I saw his fore-feet slip; he drew back immediately, and seemed for a moment at a loss how to get on. Close to these planks a cart-load of sand had been placed; he put his fore-feet on this, and looked wistfully to the other side of the drain. The boy who attends this horse, and who had gone round by another path, seeing him stand there, called him. The horse immediately turned round, and set about scraping the sand most vigorously, first with one foot then the other. The boy, perhaps wondering what he would be at, waited to see. When the planks were completely covered with sand, the horse turned round again, and unhesitatingly walked over, and trotted up to his stable and driver.'

CHAPTER XXI.

FOES OF ANIMALS.

' Oh ! 'tis excellent
To have a giant's strength ; but it is tyrannous
To use it like a giant.'—SHAKESPEARE.

'Yet pity for a horse o'erdriven,
   And love to which my hound has part
   Can hang no weight upon my heart
In its assumption up to heaven.'

In Memoriam.

AFTER the description we have given in the preceding chapters of the life, wonderful structure, intelligence, and uses of animals, we regret that necessity and duty compel us to show that increased and unceasing efforts are necessary to ameliorate the condition of those animals who suffer from man's cruelty, as well as to encourage the exercise of humanity towards them.

Although we have mentioned many different kinds of animals, from the tiny insect to our noble and valuable friend the horse, there is hardly a tribe of them but what is in some form and degree exposed to cruel treatment by man, who proudly considers himself to be not only their superior, but 'lord of the creatures.'

Eloquently and strongly as the claims of animals have been enforced by different writers and speakers, men's hearts, in many cases, are still vicious and unfeeling towards the brute creation, and humanity has to continue to wage an unceasing war against every form of cruelty.

As it would occupy too much space to define the multifa-

rious forms of cruelty to animals, we may remark that it consists in the infliction of agony and pain for its own sake, sheer
mischief and wickedness, without one iota of benefit to be
derived therefrom. Cruelty to any form of animal life, whether
it be for gain, amusement, the gratification of the appetite, the
promotion of science, or even the support of human life, is the
needless infliction of pain.

As we have already, in a previous volume, referred to a
defective education in humanity and thoughtlessness being
prolific causes of cruelty, we may now add another, namely,
the too common idea entertained by ignorant people that
animals cannot feel so acutely as human beings do, and that
there is neither harm nor cruelty in treating them roughly.
We need not say that this is nothing less than a wicked error.
It has been justly observed that, more or less, all animals give
forth signs of suffering, not only of a physical, but of a mental
kind. If the dog that is kicked gives a proof of bodily sensibility by a terrific yell, the lion, the tiger, and the panther,
robbed of their young, will give a proof of their mental agony
by making the woods and wilds echo and re-echo with their
frightful roaring. In some way or other all animals show their
sensibility to pain.

DIFFERENT KINDS OF CRUELTY TO ANIMALS.—There can
be no doubt that horses, elephants, bears, monkeys, dogs, cats,
birds, and other animals that are trained to perform in various
places, undergo, previous to the public exhibition of them, a
rigid discipline and tortures of a greater or less degree.

Whether the practices we are about to allude to are traceable to ignorance or wilful intention, those who indulge in
them are the foes of animals. The practices we refer to are—

Putting crabs and lobsters into cold water, in which they
have to remain until it boils. *Skinning eels*, and cleaning the
insides of fish before life is extinct.

Hobbling donkeys when exhausted by labour, and, thus
manacled, turning them out to graze, which they do with the
greatest difficulty.

Shearing sheep too early, and exposing them to the bitter
weather of a late winter, by which they often suffer severely.

Plucking fowls alive ; tying the legs of several of them together, and in this way carrying them to market, often head
downwards ; crowding ducks and poultry into crates or baskets,

so that they are sometimes found dead at the end of a railway journey.

Cropping the tails and ears of dogs for fashion's sake ; muzzling them in hot weather, and keeping them always chained to the kennel.

Leaving cats in empty houses, and so thereby not only depriving them of liberty, but of food and water.

Turning animals into fields that yield but little and inferior food, and leaving them there during the cold nights of winter, without a shed or sufficient shelter of any kind to screen them; and also, when stabled, neglecting to provide them with a sufficient allowance of proper food, merely from mercenary motives.

Overstocking cows intended for sale, so that their enlarged udders may induce buyers to offer higher prices for these animals, with an idea that they are good milkers.

Overcrowding vessels and railway trucks used for the transit of animals, to save expense.

The continuous confinement of *horses in mines*, by which their eyesight is often seriously injured, and sometimes entirely destroyed.

Goading animals when driving them along the road, and striking them with heavy cudgels across their legs and horns, to increase their speed.

Tormenting hedgehogs by throwing them, rolled up like a ball, into water, in order to see them open their bodies and swim to save their lives.

Setting toothed steel traps to catch hares, rabbits, or other animals by their legs, and often subjecting them, during a long cold winter night, to the severest suffering. ' It is a matter for regret with all humane persons that some measures cannot be taken to prevent the cruelties inflicted in the trapping of hares and rabbits. The aggregate of suffering thus caused is no doubt much greater than the pain which arises from pigeon-shooting, coursing, hunting with harriers, and all questionable sports put together.'

Shooting parent birds during the breeding season, by which their young suffer starvation and die a lingering death.

Never giving to *caged birds* the liberty of flying now and then round the room, so that they may enjoy the luxury of using their wings, given them by nature for that purpose, and which is their instinct and impulse to do.

When fishing with rod and line, hooking one living animal to catch another, which custom inflicts a twofold cruelty, by a deliberate crucifixion of the bait, and sometimes tearing out the gullet of the other victim. Such a system is cruelty intensified—vivisection of the worst kind; and also a wanton destruction and prodigal waste, in many cases, of the living treasures of our lakes, rivers, and streams.

Hiring donkeys and ponies, often in very poor condition, at places much frequented at holiday times, for so much the half-hour or the hour, conditionally that they are made to go as fast as possible, which, in addition to often being overladen, involves much cudgelling and ill-treatment in other ways, in order that the promised payment may be secured. In this matter the owners are not always exclusively to blame.

Exposing the same kind of animals and goats, at our wateringplaces, for many consecutive hours to the intense rays of the sun, without shelter, and often without a proper supply of food and water. Loading the carriages drawn by goats to such an extent, that in toiling on and on these animals often fall down exhausted, then are cruelly beaten because they have failed in their task. We fear many of the visitors to our wateringplaces, of whom more humanity and kindly feeling might be expected, are guilty, if not intentionally, of adding to the miseries of the animals referred to.

Slowly killing calves by bleeding them at intervals, so that their flesh, being thereby made white, may appear more tempting to the appetite, and a greater delicacy at the table.

Opening, for the same reasons, a vein under the tongues of *turkeys and other poultry,* then hanging them up by the legs so that their life may ebb out of them by degrees, or drop by drop, which often requires many hours of suffering to accomplish.

Employing men sometimes possessing the most brutal instincts, without proper training, wholly ignorant and unscientific, as *slaughterers,* some of whom often subject the animals they are employed to kill to much unnecessary pain, simply because, by their bungling use of the poleaxe, they have to give several heavy blows when one, or at most two, should suffice, if properly directed.

Cutting off the horns of cattle is a reprehensible practice, and has been pronounced by eminent veterinary surgeons to be a cruel one, and a source of great pain to the animals thus mutilated.

DRINK AND CRUELTY.—There can be no doubt that much cruelty is inflicted upon animals when those who have the care of them are excited by strong drink. If a man has not been properly educated in his duty to the lower animals, although in his sober moments he may treat them kindly enough, yet when his passions are inflamed by drink he becomes, in many cases, cruel to the brute under his charge; he cares neither for the law, the expostulations of the humane, nor for the punishment which may overtake him for his brutality.

These remarks apply specially to certain classes of men. In many instances, too, dealers and small jobbers, who in twos, threes, and fours may be seen in carts each drawn by a half-starved horse, going miles out of town on business, and who, when it is over, repair to some public-house to indulge in ' beer and 'bacca,' where they often remain until a late hour, and then whip and drive the poor animal at such a rate that, before the journey home is finished, he is ready to drop through sheer exhaustion ; and we fear that in many cases, when he reaches home, the poor jaded brute receives but scant food and attention to compensate him for his long and toil-some journey.

We are glad, however, in being able to state that, against this dark picture, we can place a bright one. There are many dealers who are not only kind-hearted, but have become noted for their humanity to the animals of which they have the charge.

SUNDAY PLEASURE-VANS.—That cruelty should be inflicted upon horses employed in Sunday pleasure-vans is much to be deplored, for more reasons than one. It is well known that the animals referred to are engaged in dragging crowded vans of pleasure-seekers, who leave London and other large towns in hundreds on the morning of the Sabbath, for journeys of ten, fifteen, and even twenty miles into the country.

In many, if not in all these cases, the horses so employed are those who have been toiling hard all the week, many of them already weary and exhausted, and by no means either well fed or properly cared for.

The journeys out are often performed under a burning sun, and but little rest and refreshment are allowed to the animals until the destinations are reached. After a few hours they have to return home. The vans are now laden with men and women often noisy and excited when they start, and who become more so by stopping to drink at the public-houses by the way. To make up for time thus wasted, the poor animals are, in many instances, severely flogged and made to go at a greatly increased speed, and often at a late hour reach home almost out of breath, their bodies steaming with sweat, and all but worn-out with the heat and toil of the long journeys they have performed. We fear God has a reckoning to make with all such Sabbath desecrators, who obtain money and pleasure at the cost of so much animal suffering.

CRUEL SPORTS.—We should imagine that all students of the rights of animals, and all those whose hearts have within them anything like humanity, will readily endorse the following quotation :

> ‘ Detested sport,
> That owes its pleasure to another's pain.’

That man has a right to hunt and to capture animals to supply the needs of his physical nature, may be admitted, but his *right to hunt them for mere sport* is very doubtful. We fear the love of sport is a very prolific cause of animal suffering. Hunting the fox, the stag, or the hare may be a very agreeable and healthy pastime for those gentlemen who indulge in it, but it is a widely different thing to the animals that are pursued.

Cunning as the fox may be, and much as he is disposed to appropriate to himself food in the shape of game and farm-yard luxuries to which some would say he has no moral right, he is nevertheless an intelligent animal, and possesses those emotional qualities which are more or less characteristic of all animals, especially of quadrupeds.

In being hunted the fox has only his own speed, strength, and stratagems to depend upon to ward off a premature and cruel death. It is a poor justification of this sport for the lovers of it to say that the animal has a wide range of country in which to run, or that it is necessary such destructive animals should be destroyed so that they may not become too numerous.

But is nothing to be said for the terror which must seize a fox when he hears behind him a pack of hungry dogs, who in a few minutes may tear his body to pieces while yet alive ?

17—2

Neither speed nor cunning can ward off the terror which the sounding horn, the cry of 'Tally-ho!' and the yelping of the hounds will produce. These remarks are equally applicable to the stag and the hare, who are hunted in a similar manner. What is there manly or courageous in a number of men, on as many horses, and assisted by a pack of hounds, following such little, defenceless, and comparatively powerless animals as a fox or a hare, and then, when the chase is over, to stand and deliberately witness the tearing and devouring of the body which had to succumb to superior strength and speed? One writer, referring to this cruel pastime, says : ' Manly, forsooth ! Call it a wild passion, a brutal propensity, or anything that indicates its nature ; but to give it any connection with reason is like making an union between black and white.'

PIGEON MATCHES.—The love of sport is insatiable. It finds victims among animals more defenceless than those already referred to. For mere wanton sport many a human savage has waited and watched for hours together, gun in hand, in order to bring down a swallow on the wing, a lark rising heavenwards, a thrush, a blackbird, or a linnet, probably hard at work in gathering materials for their nests or food for their young.

' *Pigeon-shooting*,' says Professor Wilson, ' in which half a hundred, or half a thousand, pigeons are killed merely to show the dexterity of some "crack shot," and to make money by wagers on the numbers knocked dead or mutilated, is at once scandalous to a civilized country and totally repulsive to humanity.'

It is well-known that many of the pigeons shot at during these shooting matches are mutilated only, not killed at once : it may be a leg or a wing is broken, or partially disabled, but not sufficiently to prevent the wounded creatures from flying beyond reach, when they fall in some place where they cannot be found, and so have to remain in a suffering and helpless condition until death ends their lives. Nor is this all. In many of the preliminary preparations for pigeon-shooting matches the most atrocious cruelties are practised upon the poor birds, which are supposed, by placing them in disadvantageous circumstances, to be for the success and benefit of those who are interested in the slaughter of these feathered victims.

In corroboration of our assertions we quote the following from the *Evening Standard* newspaper, of February 16, 1883 : A publican, named Ludlam, and a trapper, named Rogers, were brought up at the Belper Petty Sessions yesterday on a charge of barbarity at a pigeon match. The story would be sufficient to secure the passage of Mr. Anderson's Bill had any doubt previously existed on the subject. At a match held at Crich, in Derbyshire, these ruffians were observed to handle the birds in a peculiar way, and it was afterwards found that they had damaged the eyes of their victims, either with acid or with some sharp instrument, and had thrust pins into the flesh of the birds under the tail feathers. They were found guilty of torturing three pigeons, and most properly committed to prison for two months with hard labour, without the option of a fine. Such proceedings at pigeon matches are exceptional, if only because the opponents of the competitor so favoured would see and object; but that such things should ever be done strengthens the plea for the abolition of the ' sport.'

If sport of this kind is wicked, cruel, and unmanly, is it not particularly unwomanly and degrading to the female character to take any part in such pastimes? To those who do so we would say :

> ' If the rougher sex by their fierce sports
> Are hurried wild, let not such horrid joy
> E'er stain the bosom of the British fair.'

Can anything be more degrading to any being boasting of reasoning powers of birth, education, and position, than making amusement out of the agonies and death of any animal, especially out of those of a poor defenceless pigeon ? We do not hesitate to say that the patrons, patronesses, and supporters of pigeon-shooting clubs are guilty of the meanest, most despicable and worst kind of cruelty ; that they are setting an example calculated to demoralize, not only themselves, but society at large, and at the same time give a licence to men of brutal passions to indulge in the worst species of cruelty and crime.

While many persons seek pleasure in only witnessing these sports, their presence is nothing less than a tacit approval and encouragement of them. There can be no doubt that gambling or money-making has more to do with the practice than the aim and ambition to excel as good shots. This practice had its rise among Londoners, who, it is said, began with the sparrow, and now take the pigeon. The practice, to say the least of it, is a cowardly pastime. Fancy taking doves even for this sport ; a bird that is the emblem of peace, and of all other birds the mildest and most affectionate ; the bird that is the symbol of simplicity, innocence, and fidelity ; the bearer of the olive-branch ; the herald of safety when the waters began to recede after the deluge, and in the form of which the Spirit of God alighted on the Saviour. To make this bird the object of cruel sport is unmanly, inhuman, and unchristian.

In addition to what we have already said on pigeon matches we give the following graphic description of the cruelties of trap pigeon-shooting which appears in the *Animal World* for March, 1883. It will not only be perused with painful interest, but will show how ready the editor of the above journal is to lend his aid in suppressing such a cruel and ignoble pastime as that of

TRAP PIGEON SHOOTING.—The editor says, We have asked Mr. Harrison Weir to supply us, for the twentieth time, with an illustration against this ignoble sport, and as one of the men best acquainted with birds, to make a few remarks on the subject. Mr. Weir commences as follows : ' I appeal to the Royal Society for the Prevention of Cruelty to Animals, to all humane persons, to all who hate wanton cruelty in any form, to do their utmost to stop the wickedness of the sport of pigeon-shooting from the trap.' As far as we are concerned his appeal will not be in vain ; for our columns every year have been charged with uncompromising denunciations against that dastardly and demoralizing pastime, and we have no intention to give up the crusade. Nothing can exceed our detestation of trap-shooting, for which reason we are anxious above all things to attack it on fair issues. Without this care the battle will be lost. Its intrinsic unmanliness, its mockery of all that real sportsmen hold dear, its gambling associations, its levity as regards animal destruction, and its tendency to make human hearts callous to sympathy with the beautiful creatures, its victims, should be never absent from our thoughts when attacking the vice. Mr. H. Weir continues as follows : ' Hit—yes, grievously hit ! and yet the poor pigeon struggles, on and on, away o'er green fields, and anon o'er houses, with

happy children in the gardens at play, who look up, wondering why the beautiful bird flies so slowly; yet on and on it
goes, till at last, faint, hungry, and weary, it sinks to rest and
to suffer—no one near to note its mute agony.   Again it strives
to rise: it is too weak, it is cold, it is hungry; it struggles
feebly, it dies.   Then it is a poor jaded thing: torn flesh,
broken bones, and ruffled feathers—a useless dead thing, a
victim of sport.   Is it manly sport?   Not so very long ago a
wounded bird joined my own—a mangled creature, with a
broken leg and with its tail pulled out.   And now another has
come " grievously hit."   Yes, grievously ! shot-marks along its
back.   Shot has torn off the feathers and laid the back bare
and bleeding.   Its tail, too, like the former, is pulled out (not
cut off), so that the sporting man may find the more difficulty
in the killing it at so many yards' rise from his gun.   Fine
sport this for men—for civilized men, educated men !   Fine
sport—truly, it must be delightful sport !   And yet, and so
'tis thus the sport goes on day by day.   Beautiful, bright,
innocent pigeons are conveyed in baskets and brought out as
living targets for sport.   Let every humane person help to stay
and stop these hideous pastimes.   Other countries are moving
in the matter, are we to be the last ?   Are we less civilized than
they ?   Are we less tender-hearted, or are we more brutal?
Why is it ?   Why is such a wicked, cruel sport made of a live
plaything ?   Why must the target be alive ?—a suffering,
quivering, agonized thing if hit.   Will anyone tell me why this
is necessary ?   I ask everybody to endeavour to stop these
cruel butcheries.'   These remarks are true enough, only they
apply to all shooting at birds, whether from traps or otherwise.
It is a pity that Mr. Weir does not see that the real point
against trap-shooting is not touched by referring to the sufferings of a wounded bird.   Our antagonists are far too clever
not to retort on us that equal tortures are caused in the field
—indeed, with greater probability; because at Hurlingham,
should the bird be missed as it rises from the trap, twenty
guns of outsiders discharge at it, and few wounded birds
escape, while thousands every day die a lingering death,
during late autumn, which have been wounded, but not killed,
by field sportsmen.

BEARING REINS.—The pride and love of display have caused
many persons to be the foes of animals.   The evils and use

of the bearing-rein on horses are, at the present time, occupy
ing much public attention, and exciting considerable interest.
That the bearing-rein is not, as now used on some carriage-
horses, in any way an advantage to them, but an injury, there
can be no doubt. This is so clearly proved by papers on
this subject, issued by the Royal Society for the Prevention
of Cruelty to Animals ; in a work written by G. Fleming,
M.R.C.V.S. ; and also in a pamphlet by E. F. Flower,
on 'Bits and Bearing-reins,' that further argument on this
subject seems almost unnecessary.

The Wrong Way.

We may, however, observe that the head and tail are
important working powers of the horse, and that anything
which hinders, or restricts the freedom of their action, produces
consequences of the most injurious kind. The bit, to which
is attached a rein tightly drawn up, puts an unnecessary
pressure upon the sides of the mouth, thereby causing, if not
sores, a painful tenderness in those parts, and the head and
neck of the horse are thrown into an unnatural position.
   In the course of time the larynx of the horse becomes

affected, inflamed, and partially diseased, and finally the animal becomes a 'roarer.' He cannot properly inflate his lungs with atmospheric air because the air-passages are contracted, which no doubt affects the circulation of the blood, and, in some cases, produces apoplexy. His eyes grow dim, his sight fails, his health is affected, his usefulness is lessened, and in many instances his days are shortened.

That extreme gagging or reining up of a horse's head must be a source of pain to him is evident from his restlessness and

The Right Way.

fruitless efforts to stretch out his head sufficiently to obtain relief. A man can walk much more easily with his arms swinging to and fro than he could if they were fastened to his sides. To fasten a horse's head to his tail to keep him from stumbling seems to be a great absurdity.

That tight bearing-reins are not necessary is seen in the fact that our omnibus and cab horses do their work better without them than they used to do with them. The rejection of these vile instruments of cruelty by many private families and by some of the drivers of the 'four-in-hand' club shows that they are not only not necessary, but that they believe their use interferes both with the strength and comfort of their animals.

To the above we may add, that hundreds of veterinary surgeons condemn the use of the tight bearing-rein as being not only unnecessary but injurious in the particulars we have stated.

Lord Portsmouth once said, 'I never allow a bearing-rein to be used in my establishment, nor did my father before me; I am sure they are both useless and cruel.'

Cracknell, the four-in-hand driver, would not use a bearing-rein, although he often drove from London to Oxford in one day and back the next, including stoppages, at the rate of ten miles an hour, and was always exact to time.

SOME BRISTOL CARTERS AND THEIR HORSES.—A few years since, we delivered a lecture in Bristol on kindness to animals to an audience mainly composed of about 500 men who had the care of horses. The men were apparently interested, and paid great attention to what was said. Twelve months afterwards we gave a second lecture in the same town, and had about an equal number of listeners as on the first occasion. A novel and unexpected sight presented itself to our notice. Just before the platform several seats were occupied by not less than fifty men, all dressed in their Sunday suits; each man wearing a rosette in the buttonhole of his coat?

'What does this mean?' we inquired of one of the lady promoters of the meeting. 'Oh,' she replied, 'you will be gratified to hear that these men were convinced by your remarks last year on the evils of bearing-reins that they were doing wrong in using them, and so, like sensible men, they determined to give them up. They have done so, and their testimony is that their horses now do their work and draw heavy loads with far greater ease than they did when tightly reined up.'

These men who, in spite of much opposition and jeers from other carters, discontinued to use tight reins on their horses deserve great commendation, while what they did teaches a humane lesson, and gives considerable force to every argument advanced against bearing-reins, which in all cases, or nearly so, are needless, and in very many instances cause great suffering to our noble friend the horse.

CHAPTER XXII.

FRIENDS OF ANIMALS.

'Old Martin loved his dogs as though
  They had his children been;
And they loved him; for love can make
  E'en man and beast akin.'

ALTHOUGH, as already stated, many animals are subjected to much cruelty in order to afford sport and pastine to those who pay no regard to the laws of humanity, there are, nevertheless, signs that better and happier times are in store for them, which may well give encouragement to those who are the friends of the brute creation.

Taking into consideration the gradual change now taking place in the feelings of the public toward animals, and in their views as to the rights and claims of these dumb creatures, we may confidently look forward to the time when kindness will prevail, and shelter beneath her outspread wings every member of the animal kingdom.

These changes for the better in the general condition of animals are not, however, to be regarded as favours conferred upon them by man, but as their own chartered rights, given to them by their beneficent Creator.

INDIVIDUAL FRIENDS.

Amongst the friends of animals may be reckoned all those who have made, and are making, in their own way, individual efforts to protect them from ill-usage, and to contribute to the pleasure of their lives. Although half a century ago the idea of legislating for animals, or organizing any efforts to protect them, was ridiculed and laughed at as absurd and utopian, a galaxy of humane men and women have denounced cruelty to animals as a disgrace to human nature, to civilization and Christianity.

Going no farther back than 1809, and coming on to later years, we have the names of Erskine, Burdett, Buxton, Martin, Wilberforce, Fry, Gurney, Lushington, Broome, and a host of others who were pioneers in the cause of humanity, and who, through much opposition, nobly advocated the claims and rights of animals.

Although as time has rolled on death has removed one by one these noble friends, others have arisen to supply their places, to do their work ; and animals are still favoured with a succession of sincere and zealous friends.

Without attempting to enumerate the many and different acts of cruelty which are committed, we may notice that some humane people regard one particular kind of cruelty as being more atrocious than any other, and therefore make special individual efforts to put a stop to it. For instance, as before intimated, one friend particularly notices the cruelties of the gag bearing-rein, and makes strenuous efforts to abolish its use. Others, who are lovers of horses, seek to effect a more rational and humane method of shoeing these, our useful, hardworking animals. This being the case, we may clearly see that this is to the great advantage of animals, especially in those places where there is no society established, or organized means in operation for the suppression of cruelty.

COLLECTIVE FRIENDS.

These refer more especially to the public agencies and societies which exist for, and are employed in, the great work of humanity to the lower animals. First, we may mention the 'Royal Society for the Prevention of Cruelty to Animals,' which claims priority, and whose influence has always been of the most wonderful kind.

The following conveys much useful information respecting the origin, work, objects, and progress of the Society to 1872.

' Prior to the enactment of the statute intituled " An Act to Prevent the Cruel and Improper Treatment of Cattle " (July, 1822), introduced by the late highly esteemed Mr. Martin, the management of animals in this country was inhuman. Whether from ignorance, thoughtlessness, heedlessness or

A Group of Animals.

wanton brutality, they were subjected to extreme pain and torture, and their condition failed to excite the commiseration of the public. The most reckless and savage punishment, and the most disgusting disregard to the bodily sufferings of animals, were exhibited unconcealed in the highways and streets daily; festering sores, discharging wounds, excruciating lameness, and tottering infirmity, called not forth modern devices to evade public reprobation, and without disguise the lash and goad worked their bloody inflictions. The uncombined efforts of a few benevolent individuals were no check to these evils; and hence it became necessary to establish a society which should assemble and unite the friends of dumb animal creatures.

'The founders of this Society met on the 16th of June, 1824, and inaugurated the Society for the Prevention of Cruelty to Animals, appointed a committee, and conceived the following plan of operations:

' 1. The circulation of suitable tracts gratuitously, or by cheap sale, particularly among persons intrusted with cattle, such as coachmen, carters, and drovers.

' 2. The introduction into schools of books calculated to impress on youth the duty of humanity to inferior animals.

' 3. Frequent appeals to the public, through the press, awakening more general attention to a subject so interesting, though too much neglected.

' 4. The periodical delivery of discourses from the pulpit.

' 5. The employment of constables in the markets and streets; and

' 6. The prosecution of persons guilty of flagrant acts of cruelty, with publicity to the proceedings, and announcement of results.

' Steadily working by the above means, bravely bearing contumely, and overcoming difficulties, the founders became stronger year after year; subscribers and co-workers gradually joined the ranks, and a marked improvement slowly manifested itself in the treatment of animals. Then followed the distinguished patronage of Her Most Gracious Majesty the Queen, Her Royal Highness the late Duchess of Kent, the nobility, and many distinguished members of both Houses of Parliament; and in 1840, by command of Her Majesty, the Society was honoured with the prefix of " Royal." Since that period its progress has been regular, and its achievements encouraging, and now it is regarded as a permanently established institution, which has outlived ridicule, and secured for its founders the esteem of good and practical men of this and succeeding generations.

'Branch and sister societies are in active operation in many of the largest towns in England, Ireland, and Scotland; and more than sixty well-constituted and energetic associations of a like character are labouring in various parts of Europe, America, the Colonies, and distant parts of the globe.

'During many years the committee advocated the removal or enlargement of Smithfield Market. To the co-operation of conductors of daily journals the committee are indebted for their ultimate triumph over long-cherished prejudices and vested interests. Owing to its agency, bull-baiting, bull-running, cock-fighting, badger-baiting, and other wicked sports of a barbarous age have been prohibited by legal enactments. In 1835 the Society obtained an amendment of Martin's Act; in 1845 an amendment of the law for regulating Knackers' Yards; in 1849 a new and much improved Act for the more effectual Prevention of Cruelty to Animals; and in 1854 an Act prohibiting the use of dogs as beasts of draught or burden throughout England. Thus, the whole police force is enlisted for the protection of domestic animals.'

To show the progress of this great work since 1872, we may state that the Royal Society for the Prevention of Cruelty to Animals is extending its operations by forming new branches and employing men as officers stationed at them. Every year very many thousands of tracts on humane literature are distributed broadcast. Its own monthly organs, the *Animal World* and *Band of Mercy*, are widely circulated. A small book, containing a digest of Acts of Parliament passed in favour of animals, and many useful suggestions, is supplied gratuitously to policemen and to others who may apply for it, while sermons and lectures on the claims of animals are now given in every part of the kingdom.

It has also succeeded in obtaining a law for the protection of all wild birds during the close season, and is now actively trying to secure legal protection to ALL wild animals, not only for a part of the year, but continuously.

Independently of the assistance it renders to the police and private individuals who take proceedings against those guilty of acts of cruelty they may have witnessed, the number of convictions obtained by its own inspectors increases every year, so that the Society is becoming more and more the defender of animals and the advocate of their rights.

To be successful in obtaining convictions against all cases taken into court, thorough acquaintance with the law, a careful analysis of evidence, and great vigilance are required. We may here state that so much attention is paid by the Council and Secretary to the points above mentioned that notwithstanding the opposition often given to the Society's preventive measures, it does not lose, on an average, more than three per cent. of the cases taken before and tried by our magistrates.

Although the moiety of costs legally belongs to the Society, it never takes it, but the same is left in the hands of the presiding magistrate, to be given by him to some local charity or other object he may think proper. This being the case, the Society cannot be charged with taking cases of cruelty into court as a pretext for getting money ; on the contrary, it will be seen that the object is one in favour of and for the benefit of poor defenceless animals.

All officers employed by the Society are forbidden, under pain of dismissal, from taking any gratuity whatever, either from plaintiffs or defendants, without the sanction of the Secretary, so that in collecting evidence respecting any cases of cruelty there may be no inducement for them to act unfairly, but justly and entirely in the interest of animals that may be the victims of ill-usage.

From personal knowledge of the way in which the Society carries out its great objects we are bound to repudiate the unfounded assertion some of its enemies have made, 'that it connives at cruelty committed by the wealthy, while the poorer classes are vindictively punished for the most trivial acts of cruelty that may be committed by them.' This is not so ; and if anyone will carefully note the reports which appear in our daily and weekly newspapers respecting charges of cruelty preferred by the Society, they must, we think, be convinced that it is no ' respecter of persons,' and that it is not at all affected in its action by any difference in the social position or influence of those who may be guilty of cruelty in whatever form it may be committed.

The following extracts from the last annual report of the Society fully corroborate our assertions. In referring to what is chiefly an aristocratic sport, namely, that of

PIGEON MATCHES, it says : ' Thirteen convictions were obtained for cruelty to pigeons at or in connection with pigeon-shooting matches. Public opinion, now so often expressed against this sport, has during late years made considerable progress in this country, and still more so in the United States, owing in great measure to the protests and attacks directed against it by your Society; and your committee therefore feel that the time has now arrived when they may make a step forward by attempting to bring the practice of shooting pigeons from traps within the purview of the statutes made for the protection of animals.

' Setting dogs to worry cats, mentioned in your committee's last report, continues to be a prevalent offence. Nineteen persons, principally young men of good social position, were convicted for this scandalous misconduct, and that number would have been increased tenfold but for the difficulty of obtaining evidence against offenders of this class, whose misdeeds are frequently reported to the Society ; and, when the evidence is obtained, of securing a conviction.'

ANIMALS IN MINES.—' On many past occasions your committee have directed your serious consideration to the tendency to ill-treat horses, ponies, and donkeys working in mines, away from the observation of the public eye. During the past year upwards of twenty convictions were obtained against persons for this kind of cruelty. Various measures have been directed by your executive to encourage a better sense of humanity in the minds of persons having charge of animals underground, it is hoped with success. Recently the Baroness Burdett Coutts took part in an important movement, the May-day procession of horses at Newcastle-on-Tyne, designed in part to accomplish this object. Considering the difficulty of detecting offenders engaged below the surface of the earth, your committee strongly recommend the bestowal of rewards (which may be done by private individuals on the nomination of managers and owners of mines), on drivers and horse-keepers who have distinguished themselves by the kindly treatment of animals committed to their care.

' During the year 1881, 4,132 convictions were obtained by

18

your Society, consisting of 2,526 for ill-treatment to horses, 142 to mules and donkeys, 122 to cattle, 19 to calves, 36 to sheep, 20 to pigs, 12 to goats, 88 to dogs, 64 to cats, 42 to fowls, 9 to ducks, 3 to geese, 16 to pigeons, and the remainder to various other animals.

'This large number of convictions will be regarded as evidence of the Society's activity rather than as proving an increase of cruelty; a remark that is strengthened by the circumstance (than which nothing can better exhibit the extended operations of your Society) that the foregoing convictions were obtained in 355 distinct courts of summary jurisdiction through England and Wales.'

HUMANE SLAUGHTERING OF ANIMALS.

'Your committee rejoice to be able to direct your attention to the formation of a society for improving the present methods of slaughtering animals intended for human food, for opposing the licensing of private slaughter-houses, and for the erection of a model slaughter-house, with a better system of drainage and a more perfect method of removing and destroying offal. Your committee have rendered hearty assistance to the new society, whose object is regarded as of the first value. . . . The following extract is made from the society's prospectus :— "The objects of this society are, briefly—(*a*) to move Parliament to establish public slaughter-houses, such as exist in the Continental capitals, and in some English and Scotch cities, as Manchester, Bradford, Liverpool, Edinburgh, and Glasgow ; (*b*)·to abolish private slaughter-houses ; (*c*) to erect in the neighbourhood of the metropolis a 'model abattoir,' where all the arrangements shall be as perfect as modern science can make them, where animals shall be killed painlessly, and proper precautions taken to ensure the healthiness of the meat supplied; (*d*)to make the inspection and superintendence of private slaughter-houses, as long as they are allowed to exist, really effectual. Already the Royal Society for the Prevention of Cruelty to Animals has done much good in the direction of calling attention to the cruelty practised in private slaughter-houses. It is felt that an association on a special basis is now needed, to embrace not only considerations of cruelty to animals, but the claims of the public health as well, and to devote itself exclusively to the whole of this large and important subject." '

ANTI-VIVISECTION.

The reader will see in the following quotation, taken from one of the publications of the Royal Society for the Prevention of Cruelty to Animals, what it has done and intends to do in the question of vivisection. In referring to the provisions of a certain bill bearing on this practice, it says : 'The country has a right to know what it is doing both as regards the number of animals it permits to be tortured for scientific purposes, and the character of the suffering caused in the administration of this statute, which gives to certificated persons immunities in the performance of painful experiments for purposes, whether justifiable or not, the enlightened public opinion of the nation has now no opportunity of determining. In the draft of a bill prepared by your committee for the prevention of all cruelty committed under the name of vivisection, and submitted by them to the Royal Commission, this publicity of the results of experimentation was provided for solely to fix on operators the responsibility of their operations ; and your committee cannot but regret that in the latest and all previous returns the wise check suggested by them has apparently been designedly prevented. . . .

'Your committee are earnestly determined to pursue their endeavours to obtain more accurate information in reference to operations performed under this Act. They invite you to assist them in unveiling the dark recesses of vivisection, and detecting unlawful operations. Besides the public eye and private informers, there are several societies established solely and purposely to prevent vivisections, who may be able possibly, by proper organization, to detect offenders. To the members of these associations, as well as to the public, your committee give assurances on the present occasion, that, while engaged in many other fields of important labour for the protection of animals from the manifold causes of suffering to which they are liable, their endeavours by all legitimate means to overtake alleged and punish real offenders, whether they belong to the professional physiologists or any other class, shall not be relaxed. They call on all persons opposed to vivisection to join them, not only in applying the provisions of the statute, but in obtaining repeal of its defects by which painful operations are performed in secret without effectual supervision and check.'

It is a matter for congratulation that not only her Majesty the Queen, but other members of the Royal family, have evinced the warmest interest in the work and success of this Society, and that many of the nobility, aristocracy, bishops, clergymen, members of Parliament and of the legal profession, medical men, officers of every grade, both in the army and navy, tutors in our universities, colleges, public and private schools, editors of magazines and of newspapers, merchants and others, both ladies and gentlemen, in almost every position of life, are its supporters, active co-workers, and earnest friends.

The introduction to the fifty-eighth report will, we think, show conclusively that we have not overdrawn the picture as to the great interest taken in the Society's work. It says : 'The fifty-eighth anniversary meeting was held on Thursday, June 29th, 1882, in St. James's Hall, the noble President, Lord Aberdare, being in the chair. The hall was crowded with members and their friends, and the galleries were filled with teachers and pupils of metropolitan schools. Her Royal Highness the Princess Beatrice, having kindly and graciously accepted the invitation of the committee to attend on this occasion, was present for the purpose of presenting about fifty principal rewards earned by scholars and teachers who had composed essays on the duty of mankind towards their lower fellow-creatures. On learning the nature and object of the competition, her Royal Highness expressed her great interest in the movement, particularly as it comprehended all schools, without regard to social or sectarian characteristics, within a radius of twelve miles of Charing Cross, and contained awards given only to writers of manuscripts who had maintained ascendency in two different successive contests—namely, within the schools respectively where the essays were written, and afterwards in a competition of school against school. As this is the first occasion on which her Royal Highness has appeared in public, the members of the R.S.P.C.A. entertain a deep sense of the honour conferred on their cause by the Princess Beatrice, which they desire to acknowledge with grateful thanks. Earl Sydney (Lord Steward of her Majesty's Household), the Baroness and Mr. Burdett-Coutts, Countess of Desart, Marquis of Hertford, Lady Augusta Powlett, Lady Burdett, Lady Wolseley, Sir Lewis and Lady Pelly, Lady Garvah, Lady Burghley, Lady Peel, Lady Keppel, the Hon.

Miss Bruce, the Hon. Miss Sugden, Lady Vincent, General
Eyre, Lady Henderson, General Mills, Bishop of Gloucester
and Bristol, and Mrs. and Miss Ellicott, Cardinal Manning,
Rev. Canon Duckworth, Rev. Canon Barry, Sir Walter Sterling, Bart., Right Hon. W. E. Forster, M.P., and many other
influential ladies and gentlemen, occupied the platform.'

We may now observe that in Ireland, Scotland, and Wales
similar efforts to those referred to have been made, although
in a more limited degree. In these countries are societies and
branches with committees of ladies and gentlemen, and other
co-helpers, all alike engaged in this work of humanity. Members of the different anti-vivisection organizations may justly
be reckoned as friends of animals, as their object is not only
to lessen the sufferings of these creatures, but to prevent the
infliction of pain, which, in the practice of vivisection, they
believe is unavoidably given.

From the parent Society in London kindred ones have
sprung up, each doing its quota of good, especially to our
domestic animals. We may mention, first,

THE DOG'S HOME,

established for the avowed object of lessening the sufferings
of lost, starving, and homeless dogs, many thousands of
which have had in this home protection, proper food, and,
in all other respects, humane treatment. In hundreds of
cases lost dogs have been restored to their old homes,
friends, and owners, while in many instances those un-
claimed, if diseased, have been mercifully destroyed, and their
bodies interred in a place selected for that purpose, twenty
miles from London. Recently a meeting was held to consider
the desirability of establishing, in connection with the Dog's
Home, one for lost and starving cats, to be conducted on the
same plan as the former. An appeal has been made to the
benevolent towards this object, to which some have already
responded. All those who are acquainted with the neglect
and ill-treatment to which cats are often subjected, will, we
think, not only approve of this laudable intention, but will give
it all the support they possibly can.

Another society which has been of vast service to animals,
not only in London, but in many provincial towns, is the

METROPOLITAN DRINKING FOUNTAIN AND CATTLE
TROUGH ASSOCIATION,

whose object is not only praiseworthy and humane, but which
deserves the sympathies and support of everyone, especially
of those who own animals and are more directly benefitted
by their services. Let any person try to realize the great
pain occasioned by extreme thirst, especially during the intense
heat of summer; he will then see what a boon the vast
number of troughs and fountains, erected by this association,
must be to the hundreds of horses, sheep, oxen, and dogs,
who drink from them in the course of one day.

We have often watched animals heated by the summer sun
and parched with thirst drinking at these troughs, and as
frequently noticed a kind of grateful expression of feeling of
all of them after taking the cooling draught so much needed
but so refreshing ; and they have seemed practically to repeat
a verse of a poem we wrote on water many years ago :

> Not one gift of Nature
> Water can excel :
> Nature is its brewer,
> And she brews it well.

This association has been the means of erecting, during its
twenty-three years' existence, 497 fountains for *human beings*,
and 502 troughs for *animals*, at which it is estimated that there
are more than 250,000,000 drinkers annually. And yet there
is need for more.

THE BROWN ANIMAL SANATORY INSTITUTION.

Among the friends of animals we may justly include the
founder of the above institution, respecting which we gather
the following information from the 'Parlour Menagerie :' 'The
buildings denominated the "Brown Institution," situated
close to the Vauxhall Station of the South Western Railway,
were opened on December 2nd, 1871, and accommodation is
provided for horses, horned cattle, sheep, dogs, etc., besides
pens for poultry, and an aviary for birds. The institution had
its origin in the benevolence of the late Thomas Brown, of
Dublin, who, about twenty years previously, bequeathed the
residue of his personal estate to the Senate of the University
of London, for the founding, establishing, and upholding an

institution for investigating, studying, and, without charge beyond immediate expenses, endeavouring to cure maladies, distempers, and injuries any quadrupeds or birds useful to man may be found subject to, such institution to be under the direction of the senate; a professor, or superintendent, to be appointed by the senate, with a residence and salary, who shall give at least five lectures annually free to the public. The testator further expressed the desire that kindness to the animals received should be a general principle of the institution. He also directed, that, in order to render the endowment sufficient, the interest of the sum bequeathed should accumulate and be added to the principal during a period of fifteen years. The will being contested, a considerable period was spent in litigation; but in 1858 it was ultimately decided that the bequest and its accumulations, then amounting to nearly £23,000, should be transferred to the management of the University of London. After this decision, however, it was discovered that it was illegal to devote any part of the fund to the purchase of a freehold, while on such a site alone could the institution be legally founded. Various schemes were proposed to get over this, but without success. At length all difficulties were surmounted through the liberality of a gentleman, who purchased for nearly £3,000 a site with buildings attached. By the time the institution was opened Mr. Brown's bequest had accumulated to upwards of £30,000, the interest of which is available for the maintenance of the institution.'

'The number of animals brought to the Brown Institution for advice and treatment during 1882 amounted to 2,803. This total comprised, 1,448 horses, 64 donkeys and mules, 14 goats, 169 cats, 665 dogs, 443 chickens, and various smaller animals. Many of the cases were seen several times, so that the total number of visits for treatment has been 4,588. The patients admitted into the hospital of the institution numbered 172—comprising 44 horses, 106 dogs, 20 cats, and 2 goats. Of these 121 were discharged cured or convalescent; while of the remainder, 9 were relieved, 24 died, and 14 were destroyed, or discharged as incurable.'

We hardly need say that such an institution fully deserves the approval and encouragement of all lovers of animals, and especially of those who are the owners of them.

are also everywhere springing up, whose little members pledge themselves not only to refrain from cruelty, and to be kind to animals, but to prevail upon others as far as they can to follow their example. We may particularly mention the 'Band of Mercy' movement, 'Little Folks Humane Society,' 'Guilds of Humanity,' 'Dicky Bird' societies, etc., which already exist in great numbers both in London and the country, as well as on the Continent, and one recently established in Syria, where an English lady, devoted to the cause of animals, worked energetically for this purpose.

The thousands of little folks who compose these numerous organizations are severally under the care of ladies and gentlemen, who are deeply interested in the proper treatment of animals. Short addresses and lectures having special reference to the principles of humanity are given to them at their periodical meetings; and other educational means are used to instruct them in their duties to everything that has life, while, at the same time, they are encouraged in various ways in the practice of kindness.

Surely we may see in the efforts and agencies we have mentioned the true friends of animals; and in the shining forth of the light of humanity a sure precursor of happiness to all God's creatures.

Progress of Humanity Abroad.

Although the cause of humanity, as we have already shown, was, during its infancy, scorned and laughed at by its enemies, it has outlived all opposition; and, if we may use a figure of speech, it has now become a great tree deeply rooted, whose branches, like those of the banyan tree, are spreading far and wide into other lands, there taking root, and again throwing out in all directions other branches, under which animals may find shelter and protection against every form of cruelty.

Using another metaphor, Humanity, as a winged messenger, is in every quarter of the wide world lifting up her voice, to which people of almost every nation, kindred, and tongue lend a listening ear, and are responding to her pleadings for gentleness and kindness to be extended to all God's dumb creatures.

France, land of the witty, the warlike, and gay, has heard

this voice, which, passing over the snow-clad summits of the Pyrenees, has found an echo down amongst the Spaniards, whose natural characteristics are a curious mixture of pride, suspicion, and generosity.

This voice has also travelled to Portugal, the land of buried cities ; and to Switzerland, where the Alps rear their heads to heaven, where glaciers glisten in the sunshine like seas of ice, the land of dashing torrents and quiet lakes, rugged rocks and beautiful valleys, where live a people of simple manners, cleanly habits, of liberal thoughts, fair in their dealings, affectionate one to another, moral and religious, lovers of song, devoted to their country—the home of Tell, Lavater, Zimmermann, Rousseau, and Calvin.

In Italy, the land of beauty, music, sculpture, paintings, poetry, and song, this voice has been raised, and has reverberated amongst the fine old buildings of Rome, Florence, Turin, Milan, and of other towns where individual and organized efforts are being successfully made to introduce a new era of humanity to animals.

Austria and Germany have also given a welcome to this voice of mercy, and in many important towns it has been heard with marvellous effect ; laws have been passed for the special protection and benefit of animals, many preventive measures have been adopted, and several moral and educational agencies are now in operation for the same object, and in every way to improve their condition.

Norway, Denmark, Sweden, and Holland, famed not only for timber, dykes, and mills, but for fine, wild, rugged, grand and majestic scenery, can boast of men and women who are waging a war with the demon cruelty, and by constant progress in their merciful efforts are proving themselves to be the true friends of animals.

Russia, with its masses of ice and snow, with its crystal palaces not made with hands, has given heed to humanity's voice. St. Petersburg, with its grand buildings, quays, streets, and statues ; and Moscow, with its far-famed bells and churches, her towers and temples, both contain people of all climes, many of whom are not unmindful of the claims of animals. Amongst them are those of tender hearts and compassionate feelings, who, amid all the strife—political, social, and religious—existing there, and convulsing the empire from its centre to its circumference, are devoting their energies, time, money, and influence on behalf of dumb animals.

Egypt, with its many spots of hallowed ground, with its interesting and historical associations, with its giant pyramids, its buried cities, and the ever-flowing Nile, has, in some measure, responded to the appeals for mercy and kindness made to them by animals. Now, this country, so great in ancient times, encourages education, learning, commerce, and many useful branches of industry. In Alexandria and Cairo are many whose hearts are in sympathy with suffering animals, and who are quietly but perseveringly working to protect them from ill usage.

The voice of humanity is extending wider and wider. It has reached the Cape of Good 'Hope; it is heard in India, the land of sacred temples and rivers, of jewels and diamonds rare. It has crossed the great deep to Australia, Tasmania, New Zealand, and some of the islands of the sea, where it has been effectually heard, and numbers of noble hearts are obeying its dictates.

In the many states of America, from New York to distant California, societies have been formed to advocate the claims and to defend the rights of animals. Laws have been passed for this object, and are rigidly enforced by these societies and the friends of animals generally. Where Christianity lifts up her voice, humanity must be heard and practised. Past progress augurs well for the future welfare, and more humane treatment, shall we say, of animals the wide world over.

HUMANITY AND PUBLIC MEN.

It has been well observed that mercy is as much a lesson to be learned as any other maxim in moral philosophy. Children and savages are often cruel because they do not understand cruelty when they see it.

As human beings rank higher in the scale of creation and intelligence than animals do, they are bound, on this account, to treat them humanely, and to act as their protectors against ill-usage in whatever form it may present itself. Animals are given to man as a sacred trust which he can no more violate with impunity than darkness can come from the sun. Humanity to animals is every man's duty to practise individually, while the advocacy of it is more particularly the duty of public men. Of these we shall first refer to

MINISTERS OF RELIGION.

As Christianity and Humanity may be regarded as mother and daughter, ministers of religion are therefore morally bound by the principles they profess to hold and teach, to plead the cause of the brute creation. In referring to man's duty to the lower animals, and the under-current of kindness manifesting itself towards them, the Rev. Thos. Jackson says in one of his sermons on this subject :

'It might be expected that a sentiment so deeply rooted in the human breast, and so variously operating in all ages would be amply illustrated, and its proper limits clearly defined in the Holy Bible. Nor is that natural expectation disappointed. The eloquent language of Scripture everywhere melts with tenderness and compassion towards the dumb and brute creation. The mind that does not recognise the truth will miss the meaning and force of some of the most suggestive parables and delicate analogies of revelation. The miraculous speaking of Balaam's ass, for instance, "rebuked" more than one sort of "madness" in the prophet. It was an additional reason for sparing Nineveh, that doomed capital "wherein were more than six score thousands who could not discern their right hand from their left," that there were also in the city "much cattle." Our blessed Saviour proved that it was lawful to do a greater good on the Sabbath day, by reminding His accusers that they themselves did a lesser, but still a real good, when He said unto them, "What man shall there be among you that shall have one sheep, and if it fall into a pit on the Sabbath day, will he not lay hold on it, and lift it out? How much then is a man better than a sheep?" Does he wish to teach the blessed doctrine of the special Providence of our Heavenly Father? . He illustrates it by asserting that not a sparrow droops its wing without His notice. "Are not two sparrows sold for a farthing? and one of them shall not fall to the ground without your Father's knowledge." His unwearied watchfulness and infinite goodness were shown in the care He bestowed on the meanest creatures, and would be displayed far more conspicuously in the direction, the control, the present and final salvation of them that love Him.'

The inference to be deduced from the above remarks is that as the teachings of Holy Writ, and especially those of the Saviour, are characterized by the spirit of humanity, it is the duty of all Christian ministers to enforce on their hearers the practice of kindness to the lower animals.

MAGISTRATES, AND OTHERS.

Magistrates, and those armed with the power of the law, are expected to be ' a terror to evil-doers, and a praise to those that do well,' also to be dispensers of justice, upholders of right, and the protectors of the helpless and innocent. If in these things they fall short of their duty, they are not worthy either of the honour or the power with which they are invested.

LEGISLATORS.

The laws passed for the good government of a country should not only provide for the protection of the lives and property of human beings, but for the defence of animals too, because while those laws help to stem the torrent of cruelty, which has always a demoralizing effect, they would also aid all religious and humane efforts, and help to foster morality, to bring peace, and to produce a higher, nobler, and more refined feeling throughout society generally, especially amongst those who may have the care of animals. Such laws would act in many cases as a deterrent to those who might be inclined to treat animals cruelly.

TEACHERS IN PRIVATE AND PUBLIC SCHOOLS.

Tutors in schools and in private families should remember that

' As the twig is bent the tree is inclined,'

and that as the minds and hearts of the young are more impressible than those of riper years, they are bound not only to impart useful knowledge to the youth committed to their care, but to prune their hearts of evil propensities, and to ingraft in them true generosity, tenderness and compassionate feeling, and kindness to all living things. The necessity of attending to what we have just stated will be seen in the following extract from one of the tracts of the R.S.P.C. to Animals : ' When children grow up to be boys or lads they have special need to be guarded against habits of cruelty, as many of their amusements have that tendency ; such as birds-nesting, cock-throwing, and the like. As an antidote to these habits we should instil into young minds the domestic habits of these creatures, their affection for their offspring, their attachment to man when treated with gentleness, and the uses and comforts to be derived from them.'

PARENTS AND GUARDIANS OF YOUTH

will do well to consider the following remarks taken from a tract entitled 'Humanity to Animals Recommended.' It says:

' Not only is cruelty unnatural and abhorrent to the original constitution of human nature, but it is peculiarly criminal in man, considered as a sinner, whose very preservation in existence is only owing to the mercy of his Creator. Yet, strange as it may appear, this monument of mercy, from the cradle to the grave, is in innumerable instances prone to tyrannize over all the subjects in his power.

' Scarcely does the child possess the use of his fingers but he begins to torment the fly that buzzes and plays around him, and to deprive it of a leg or wing, in order to amuse himself with its lameness or its misery. When the little hero grows somewhat older, he sticks a pin through the cockchafer, and is delighted with its agonies ; and there are parents so depraved that they encourage these cruelties, as if they did not know that cruelty to animals is the direct road to cruelty to our fellow-creatures, and to its final reward—the gallows.

' Children that are not checked in one cruel diversion will naturally go to another. Sometimes it is shocking to see with what barbarity the kitten or puppy is treated by the little tyrants of the family. But children should be taught that animals have feelings as well as men ; and that a blow on the head or legs of these causes as much pain as we ourselves would receive from the like violence. And perhaps in some creatures of small and delicate contexture the pain may be exquisite in proportion as the frame is tender.

> ' " The poor beetle that we tread upon,
> In corp'ral suff'ring feels a pang as great
> As when a giant dies."—SHAKESPEARE.

' An error very common among young people is that animals are to be treated according to their beauty or deformity ; so that the pretty robin is caressed with gentleness and tenderness, while the toad, though equally innocent, is pursued to death with relentless cruelty. But if the same spirit were to grow with us in life, how lamentable would be its effects !'

MASTERS AND SERVANTS.

As 'example is stronger than precept,' those who own animals and wish them to be properly treated, should carry out the law of kindness themselves as an inducement to others to do the same.

CONSEQUENCE OF CRUELTY.

If indulged and delighted in, how it will grow and harden the feelings ! Like every other sin, it is impossible to say to what it may lead. In this way, perhaps, the murderer may at first have given loose to the violence of his feelings, and, by reason of not checking them, been led on in the current of unsubdued passion to an awful end. Cruelty should there-fore be curbed in the beginning, before the character and habit become fixed.

ROUGH USAGE AND KIND TREATMENT.

We observed one day, when at the Victoria railway station, a good-natured-looking cabman standing by his horse. Having expressed our pleasure in seeing the animal in such good condition, the owner said : ' Yes, sir, I flatter myself he is much better cared for than he used to be. At one time he belonged to a man who used him roughly, and that so spoiled his temper that he became a regular jibber, and the man could do nothing with him. The horse was then neglected, and became so poor that he seemed to be worth little or nothing. So he asked me to buy him. I did so, for seven pounds. I knew the horse only required proper treatment in feeding, grooming, and working him. He has had it, and there he is, sir, as good, useful, and willing a horse as ever comes into this station. He never jibs now. I wouldn't take thirty pounds of any man's money for him, if I had the chance to do so. So you see, sir, that after all,

'KINDNESS PAYS THE BEST.

In the first place,' continued the cabman, 'a man who treats his horse properly, has less trouble in making him go, while it renders the animal all the more ready and able to do his work ; so that there is a benefit on both sides. You may depend upon it, sir, that horses are sensible enough to know

when they are treated as they ought to be, and, just like ourselves, are affected accordingly. Then, on the other hand,
it's mean, base, and cowardly for a man to be cruel to a poor
brute that hasn't the chance of defending himself. What I've
got to say is : Treat your *horse well*, and *you'll* do *well ;* treat
him *badly*, and *you'll* do *badly.'*

If animals have their vices they are seldom cured of them by
rough usage. Horses and other animals may require correction,
but it should be tempered with mercy, and administered, not
in anger, but calmly, firmly, and with judgment. This will do
more in making animals willingly subserve our purposes than
all the severity and cruelty in the world could ever effect.

Kindness is an angel of mercy who spreads her sheltering
wings to stay the savage blow, and to lessen the tortures inflicted by the demon Cruelty. She opens the mouth of
humanity for dumb animals; is the advocate of their rights,
and her heaven-born mission is one of light, health, and
happiness to all living creatures.

As man never appears more noble and dignified than when
he sympathizes with those in suffering, and tries to soften the
sorrows and to relieve the pains of others, be they human
beings or the lower animals, we would say to our readers :

> ' Ever let your hearts be tender,
>   For the dumb and helpless plead ;
> Pitying leads to prompt relieving,
>   Kindly thought to kindly deed.'

Everything on the earth, in the heavens, in the air, and in
the deep blue sea bears the impress of God's love and kindness, and they say to us in ten thousand ways, ' Be ye therefore
merciful, as your Father also is merciful.' Unnumbered
testimonies in favour of kindness and gentleness come *now*
from our lunatic asylums, our factories, our mines, our universities, our schools, our homes, and from every living thing
in the animal kingdom ; and we may hear them in one loud,
grand, and universal voice, proclaiming to the wide earth that

> '*KINDNESS* HATH ITS VICTORIES,
> GREATER THAN THOSE OF WAR.'

THE END.

BILLING AND SONS, PRINTERS, GUILDFORD AND LONDON.

NEW, POPULAR,

AND

ILLUSTRATED BOOKS,

FOR THE

LIBRARY, PRESENTATION, &c.

PUBLISHED BY

JOHN HOGG,

13, Paternoster Row,

LONDON, E.C.

EDITION DE LUXE.

WITH STOTHARD'S ILLUSTRATIONS, ENGRAVED BY HEATH.

In one volume, demy 8vo., cloth, 15s. ; half morocco extra, gilt edges, 25s.

The Life and Adventures of Robinson

Crusoe, with a Sketch of De Foe, by HENRY J. NICOLL. (Printed from a new fount of old-faced type.)

NOTE.—This is a complete, unabridged edition of De Foe's masterpiece, with all the twenty-two beautiful Illustrations from the Drawings by THOMAS STOTHARD, R.A., engraved by CHARLES HEATH. These Illustrations are now printed from the Original Copper Plates, which were produced at great cost, and are still in perfect condition, having been steel-faced to preserve them. Copies of the Original Edition containing these plates, published by Messrs. Cadell and Davies in 1820, now fetch a high price in the auction rooms.

Manuals of Self-Culture for Young

Men and Women.

WITH SEVENTY-FIVE WOOD ENGRAVINGS.

Small crown 8vo., 288 pp., cloth, price 2s. 6d. ; gilt edges, 3s.

Facts and Phases of Animal Life, and

the Claims of Animals to Humane Treatment. Interspersed with Original and Amusing Anecdotes. By VERNON S. MORWOOD, Lecturer to the Royal Society for the Prevention of Cruelty to Animals.

CONTENTS :

" We have read parts of this work with great pleasure, and intend to go through it page by page for our own personal delectation. Two-and-sixpence will be well spent upon a book which teaches humanity to animals while it amuses the youthful reader."—*Sword and Trowel.*
" It would serve well for a gift-book."—*Guardian.*
" The peculiarities of nearly two hundred animals to be found in this country are described in a manner which is throughout entertaining."—*Dundee Advertiser.*
" This copiously illustrated little volume is crowded with useful facts and interesting anecdotes."—*Echo.*
" A decided improvement on the general run of natural histories for young people."—*Daily Chronicle.*
" Young people with a taste for natural history will be delighted with its pages, and we can strongly recommend it for either a prize or an addition to the school library."—*School Newspaper.*
" An excellent little book."—*Daily News.*
" A capital natural history book."—*Graphic.*
"Crammed with good stories."—*Sheffield Independent.*

WITH SEVENTY-EIGHT ILLUSTRATIONS.

Small crown 8vo., 288 pp., cloth, price 2s. 6d. ; gilt edges, 3s.

Far-Famed Tales from the Arabian

Nights' Entertainments. Illustrated with Seventy-eight wood Engravings, and carefully revised for Young Readers.

CONTENTS :

London : John Hogg, 13, Paternoster Row, E.C.

WITH EIGHTY-ONE ILLUSTRATIONS.

Small crown 8vo., 288 pp., cloth, price 2s. 6d. ; gilt edges, 3s.

Wonderful Animals : Working,

Domestic, and Wild. Their Structure, Habits, Homes, and Uses —Descriptive, Anecdotical, and Amusing. By VERNON S. MORWOOD, Author of "Facts and Phases of Animal Life," and Lecturer to the Royal Society for the Prevention of Cruelty to Animals.

CONTENTS :

CHAP.
1. CURIOUS ODDS AND ENDS ABOUT ANIMALS.
2. PEEPS DOWN A MICROSCOPE.
3. LILLIPUTIAN SUBJECTS OF THE ANIMAL KINGDOM.
4. INSECT ARMIES, AND HOW RECRUITED.
5. AN UNDERGROUND CITY OF LITTLE PEOPLE.
6. FISH IN ARMOUR.
7. FIRST COUSINS, OR OUR BIRDS IN BLACK.
8. FEATHERED FEEDERS ON FISH, FLESH, AND FOWL.
9. PEACEFUL MONARCHS OF THE LAKE.
10. BIPED TENANTS OF THE FARM YARD.

CHAP.
11. FOREST ACROBATS, LITTLE MARAUDERS, AND FLYING ODDITIES.
12. FEEBLE FOLK, FISHERS, AND POACHERS.
13. BRISTLY PACHYDERMS, WILD AND TAME.
14. ARISTOCRACY OF ANIMALS.
15. AN ANCIENT FAMILY.
16. LOWINGS FROM THE FIELD AND SHED.
17. FOUR-FOOTED HYBRIDS, OR HALF-AND-HALF RELATIONS.
18. OUR DONKEYS AND THEIR KINDRED.
19. EVERYBODY'S FRIEND.
20. ANECDOTES OF EVERYBODY'S FRIEND.
21. FOES OF ANIMALS.
22. FRIENDS OF ANIMALS.

WITH TWENTY-EIGHT ILLUSTRATIONS.

Small crown 8vo., 288 pp., cloth, price 2s. 6d. ; gilt edges, 3s.

The Shoes of Fortune, and other

Fairy Tales. By HANS CHRISTIAN ANDERSEN. With a Biographical Sketch of the Author, a Portrait, and Twenty-seven Illustrations by OTTO SPECKTER and others.

CONTENTS :

BIOGRAPHICAL SKETCH : HANS CHRISTIAN ANDERSEN, HIS LIFE AND GENIUS.
THE SHOES OF FORTUNE :
  I. A BEGINNING.
  II. WHAT BEFELL THE COUNCILLOR.
  III. THE WATCHMAN'S ADVENTURE.
  IV. A MOMENT OF HEAD IMPORTANCE.—AN EVENING'S "DRAMATIC READINGS."—A MOST STRANGE JOURNEY.
  V. THE METAMORPHOSIS OF THE COPYING CLERK.
  VI. THE BEST THAT THE GOLOSHES GAVE.
THE FIR-TREE.
FIVE FROM A POD.
THE STEADY TIN SOLDIER.
TWELVE BY THE POST.
THE FEARSOME UGLY DUCKLING, THAT TURNED OUT TO BE A SWAN.
THE SHEPHERDESS AND THE CHIMNEY-SWEEP.

THE SNOW-QUEEN, IN SEVEN STORIES :
  I. WHICH TREATS OF A MIRROR AND OF THE SPLINTERS.
  II. A LITTLE BOY AND A LITTLE GIRL.
  III. THE FLOWER-GARDEN.
  IV. THE PRINCE AND PRINCESS.
  V. THE LITTLE ROBBER-MAIDEN.
  VI. THE LAPLAND WOMAN AND THE FINLAND WOMAN.
  VII. IN THE PALACE OF THE SNOW-QUEEN, AND WHAT HAPPENED AFTERWARD.
THE LITTLE OCEAN-MAID.
THE ELFIN MOUND.
OLD WINK, WINK, WINK.
THE LEAP-FROG.
THE ELDER BUSH.
THE BELL.
HOLGER DANSKE.
THE EMPEROR FREDERICK BARBAROSSA.

London : John Hogg, 13, Paternoster Row, E.C.

Small crown 8vo., 472 pp., cloth, price 6s.

Landmarks of English Literature.

By HENRY J. NICOLL, Author of " Great Movements," &c.

CONTENTS:

INTRODUCTION : Explains the Plan of the Book, and gives some Hints on the Study of Literature.

THE DAWN OF ENGLISH LITERATURE.

THE ELIZABETHAN ERA.

THE SUCCESSORS OF THE ELIZABETHANS.

THE LITERATURE OF THE RESTORATION.

THE WITS OF QUEEN ANNE'S TIME.

OUR FIRST GREAT NOVELISTS.

JOHNSON AND HIS CONTEMPORARIES.

THE NEW ERA IN POETRY.

SIR WALTER SCOTT AND THE PROSE LITERA- TURE OF THE EARLY PART OF THE NINE- TEENTH CENTURY.

OUR OWN TIMES.

PERIODICALS, REVIEWS, AND ENCYCLOPÆDIAS.

" We can warmly commend this excellent manual. Mr. Nicoll is a fair and sensible critic himself, and knows how to use with skill and judgment the opinions of other critics. His book has many competitors to contend with, but will be found to hold its own with the best of them."—*St. James's Gazette.*

" Mr. Nicoll's facts are commendably accurate, and his style is perfectly devoid of pre- tentiousness, tawdriness, and mannerism, for which relief in the present day an author always deserves much thanks from his critics."—*Saturday Review.*

" Mr. Nicoll has performed his task with great tact, much literary skill, and with great critical insight. No better book could be put into the hands of one who wishes to know some- thing of our great writers, but who has not time to read their works himself ; and no better guide to the man of leisure who desires to know the best works of our best writers and to study these in a thorough manner. Mr. Nicoll's literary estimates are judicious, wise, and just in an eminent degree."—*Edinburgh Daily Review.*

" Mr. Nicoll's well-arranged volume will be of service to the student and interesting to the general reader. Biography and history are combined with criticism, so that the men are seen as well as their works. . . . . The copious and careful table of chronology gives a distinct value to the book as a work of reference. The volume is without pretension, and deserves praise for simplicity of purpose, as well as for careful workmanship."—*Spectator.*

Crown 8vo., 576 pp., cloth, price 6s. 6d. ; gilt edges, 7s.

Woman's Work and Worth in Girl-

hood, Maidenhood, and Wifehood. With Hints on Self-Culture and Chapters on the Higher Education and Employment of Women. By W. H. DAVENPORT ADAMS.

" It is a small thing to say that it is excellent, and it is only justice to add that this all-im portant subject is dealt with in a style at once masterly, erudite, charming."—*Social Notes.*

" As an aid and incitement to self-culture in girls, and pure and unexceptionable in tone, this book may be very thoroughly recommended, and deserves a wide circulation."—*English- woman's Review.*

" It is a noble record of the work of women. . . . . and one of the very best books which can be placed in the hands of a girl."—*Scholastic World.*

WITH EIGHT PORTRAITS, 464 pp., crown 8vo., cloth, price 6s.

Great Movements and those who

Achieved Them. By HENRY J. NICOLL, Author of "Landmarks of English Literature," &c.

" A useful book. . . . . Such work . . . . should always find its reward in an age too busy or too careless to search out for itself the sources of the great streams of modern civilization." —*Times.*

" An excellent series of biographies. . . . . It has the merit of bespeaking our sympathies, not as books of this class are rather apt to do, on the ground of mere success, but rather on the. higher plea of adherence to a lofty standard of duty."—*Daily News.*

" Immense benefit might be done by adopting it as a prize book for young people in the upper classes of most sorts of schools."—*School Board Chronicle.*

London : John Hogg, 13, Paternoster Row, E.C

Our Homemade Stories. By Ascott

R. HOPE, Author of " Stories of Young Adventurers," &c.

CONTENTS :

Introduction - - - -		Spinning a Story.
1. PLAYING THE FOOL -	- -	A Story of a Lady's School.
2. MY DESERT ISLAND -	- -	A Story of the Canadian Backwoods.
3. THE BLACK BOOK -	- -	A Story of a Juvenile *Cause Célèbre*.
4. CROSSING THE LINE -	-	A Story of Sea Life.
5. CAUGHT OUT	- -	A Story told in a Train.
6. A SCENE FROM HISTORY -	-	A Story of a French Revolution.
7. THE GUISARDS	- -	A Story of Scotland.
8. THE SECRET SOCIETY -	- -	A Story of School Life.
9. AT THE MASTHEAD -	- -	A Story of a Storm on Shore.
10. A NIGHT IN THE BLACK FOREST		A Story of Strange Adventure.
11. BABY BOY -	- -	A Story of the Latin Grammar.
12. THE BANSHEE	- -	A Story of Ireland.

Evenings away from Home : A

Modern Miscellany of Entertainment for Young Masters and Misses. By ASCOTT R. HOPE, Author of " Our Homemade Stories," etc.

CONTAINING, AMONG OTHER ENTERTAINING AND INSTRUCTIVE ARTICLES :

The Astonishing Adventures of Jack Robinson.
Remarkable Travels in Undiscovered Regions.
The Holiday Task, a First-class Magazine written by Juveniles for Juveniles.
The Champion Charades of the Universe.
The Trials and Travels of an Ancient Adventurer.
Sketches of Manners and Customs in Monkey Land.
Three Ghost Stories, with Moonlight and Gooseflesh Effects.
A Tale of Horror in the best style of the Penny Dreadfuls.
Early Efforts of Promising Poets.
The Knight of the Woods, a Thrilling Romance of Chivalry.
Stories of School Life, etc., by the Nine Muses.
Tales about Giants, Princesses, Pirates, Indians, Enchanters, Smugglers, etc., etc.

London : John Hogg, 13, Paternoster Row, E.C.

WITH EIGHT ILLUSTRATIONS ON TONED PAPER.

Fourth edition, small crown 8vo., 384 pp., cloth, price 3s . 6d. ; gilt edges, 4s.

The Secret of Success; or, How to

Get on in the World. With some Remarks upon True and False Success, and the Art of making the Best Use of Life. Interspersed with Numerous Examples and Anecdotes. By W. H. DAVENPORT ADAMS, Author of "Plain Living and High Thinking," etc.

"Mr. Adams's work is in some respects more practical than Mr. Smiles's. He takes his illustrations more from the world of business and commerce, and their application is unmistakeable . . . . There is much originality and power displayed in the manner in which he impresses his advice on his readers."—*Aberdeen Journal.*

"There is a healthy, honest ring in its advice, and a wise discrimination between true and false success. . . . . Many a story of success and failure helps to point its moral."—*Bradford Observer.*

"The field which Mr. Adams traverses is so rich, extensive, and interesting, that his book is calculated to impart much sound moral philosophy of a kind and in a form that will be appreciated by a large number of readers. . . . The book is otherwise a mine of anecdote relating to men who have not only got on in the world, but whose names are illustrious as benefactors to their kind."—*Dundee Advertiser.*

WITH TWO COLOURED PLATES AND EIGHT PAGE ILLUSTRATIONS.

Third edition, small crown 8vo., 400 pp., cloth, price 3s. 6d. ; gilt edges, 4s.

Our Redcoats and Bluejackets: War

Pictures on Land and Sea. Forming a Continuous Narrative of the Naval and Military History of England from the year 1793 to the Present Time, including the Afghan and Zulu Campaigns, Interspersed with Anecdotes and Accounts of Personal Service. By HENRY STEWART, Author of "Highland Regiments and their Battles," "The Romance of the Sea," etc. With a Chronological List of England's Naval and Military Engagements.

"A capital collection of graphic sketches of plucky and brilliant achievements afloat and ashore, and has, moreover, the advantage of being a succinct narrative of historical events. It is, in fact, the naval and military history of England told in a series of effective tableaux."—*The World.*

"It is not a mere collection of scraps and anecdotes about our soldiers and sailors, but a history of their principal achievements since the beginning of the war in 1793. The book has charms for others than lads."—*Scotsman.*

"Besides being a work of thrilling interest as a mere story-book, it will also be most valuable as a historical work for the young, who are far more likely to remember such interesting historical pictures than the dry lists of dates and battles which they find in their school-books. . . . . Possesses such a genuine interest as no work of fiction could surpass."—*Aberdeen Journal.*

London: John Hogg, 13, Paternoster Row, E.C.

WITH UPWARDS OF 300 ENGRAVINGS BY BEWICK AND OTHERS.

FOURTH AND CHEAP EDITION.

Large crown 8vo., 520 pp., cloth, price 3s. 6d. ; gilt edges, 4s.

The Parlour Menagerie : Wherein

are exhibited, in a Descriptive and Anecdotical form, the Habits, Resources, and Mysterious Instincts of the more Interesting Portions of the Animal Creation. Dedicated by permission to the Right Hon. the Baroness Burdett-Coutts (President) and the Members of the Ladies' Committee of the Royal Society for the Prevention of Cruelty to Animals.

WHITE EYELID MANGABEY.

Specimen of the 66 Wood Engravings by Thomas Bewick in the "Parlour Menagerie."

From Professor OWEN, C.B., F.R.S., &c. (Director, Natural History Depart., B. Museum).

To the Editor of the *Parlour Menagerie.*

" The early love of Nature, especially as manifested by the Habits and Instincts of Animals to which you refer, in your own case, is so common to a healthy boy's nature, that the *Parlour Menagerie*, a work so singularly full of interesting examples culled from so wide a range of Zoology, and so fully and beautifully illustrated cannot fail to be a favourite with the rising generation—and many succeeding ones—of Juvenile Naturalists. When I recall the ' Description of 300 Animals ' (including the Cockatrice and all Pliny's monsters) which fed my early appetite for Natural History, I can congratulate my grandchildren on being provided with so much more wholesome food through your persevering and discriminating labours.

" RICHARD OWEN."

From the Right Hon. JOHN BRIGHT, M.P.

To the Editor, *Parlour Menagerie.*

" I doubt not the *Parlour Menagerie* will prove very interesting, as indeed it has already been found to be by those of my family who have read it. I hope one of the effects of our better public education will be to create among our population a more humane disposition towards what we call the inferior animals. Much may be done by impressing on the minds of children the duty of kindness in their treatment of animals, and I hope this will not be neglected by the teachers of our schools. . . . . I feel sure what you have done will bear good fruit.

" JOHN BRIGHT."

" The *Parlour Menagerie* is well named. Full as an egg of information and most agreeable reading and engravings, where before was there such a menagerie ?"—*Animal World.*

" We have never seen a better collection of anecdotes and descriptions of animals than this, and it has the great advantage of numerous and admirable woodcuts. Pictorial illustrations form an important and valuable addition to any such collection. Those in the book before us are of remarkable excellence. . . . . We highly commend the spirit which pervades the book, a spirit intensely alien to cruelty of every kind. A great deal of care and trouble has evidently been devoted to the compilation of this book. On the whole, it is one of the very best of its kind, and we warrant both its usefulness and acceptability."—*Literary World.*

London : John Hogg, 13, Paternoster Row, E.C.

WITH EIGHT ILLUSTRATIONS ON TONED PAPER.

Second edition, small crown 8vo., 352 pp., cloth, price 3s. 6d. ; gilt edges, 4s.

Boys and their Ways : A Book for and
about Boys. By ONE WHO KNOWS THEM.

CONTENTS.

Chaps. 1. The Boy at Home.—2. The Boy at School.—3. The Boy in the Play-
ground.—4. The Boy in his Leisure Hours.—5. Bad Boys.—6. Friendships of
Boys.—7. The Boy in the Country.—8. How and What to Read.—9. Boyhood of Famous Men.—10. The Ideal Boy.

"The table of contents gives such a bill of fare as will render the boy into whose hands
this book falls eager to enjoy the feast prepared for him. . . . We venture to predict for this
charming book a popularity equal to 'Self-Help.' . . . No better gift could be put into a boy's
hands, and it will become a standard work for the school library."—*Scholastic World.*

"Who the author of this book is, has been kept a secret, and the anonymity we regret,
because the work is one with which no writer need be ashamed to identify his name and stake
his reputation."—*Edinburgh Daily Review.*

"It is a boy's book of the best style."—*Aberdeen Journal.*

WITH EIGHT PORTRAITS ON TONED PAPER.

Dedicated by permission to the Rt. Hon. W. E. GLADSTONE, *M.P., &c.*

Third edition, small crown 8vo., 384 pp., cloth, price 3s. 6d.; gilt edges, 4s.

Plain Living and High Thinking; or,
Practical Self-Culture : Moral, Mental, and Physical. By W. H.
DAVENPORT ADAMS, Author of " The Secret of Success, &c.

PART I.—MORAL SELF-CULTURE.

Chap. 1. At Home. | Chap. 3. Character.
,, 2. Life Abroad. | ,, 4. Conduct.

PART II.—MENTAL SELF-CULTURE.

Chap. 1. How to Read.

Chaps. 2 to 9. Courses of Reading in English Poetry, History, Biography, Fiction,
Travel and Discovery, Theology, Philosophy and Metaphysics, Miscellaneous
Science and Scientific Text Books. Chap. 10. How to write : English Com-
position.

PART III.—PHYSICAL SELF-CULTURE.

" Mens sana in corpore sano."

"We like the thorough way in which Mr. Adams deals with 'Self-Culture : Moral, Mental
and Physical.' His chapter on the courtesies of home life, and the true relation between
parent and child, is specially valuable nowadays. . . . He certainly answers the question,
'Is life worth living?' in a most triumphant affirmative."—*Graphic.*

"Books for young men are constantly appearing—some of them genuine, earnest, and
useful, and many of them mere products of the art of book-making. We have pleasure in
saying that this volume by Mr. Adams deserves to take its place among the best of the first-
mentioned class. It is fresh, interesting, varied, and, above all, full of common sense,
manliness, and right principle."—*Inverness Courier.*

"Young men who wish to make something of themselves should invest seven sixpences in
this most valuable volume."—*Sword and Trowel.*

"A better book of the class in all respects we have seldom had the pleasure to notice. . . .
We cannot too strongly recommend it to young men."—*Young Men's Christian Association
Monthly Notes.*

London: John Hogg, 13, Paternoster Row, E.C.

"The best book of the kind."
"A complete Society Encyclopædia." } *Vide Critical Notices.*

With Frontispiece, small crown 8vo., 352 pp., handsomely bound in cloth, price 3s. 6d. ; gilt edges, 4s.

The Glass of Fashion : A Universal
Handbook of Social Etiquette and Home Culture for Ladies and Gentlemen. With Copious and Practical Hints upon the Manners and Ceremonies of every Relation in Life—at Home, in Society, and at Court. Interspersed with Numerous Anecdotes. By the LOUNGER IN SOCIETY.

CONTENTS :

CHAP.
1. AT HOME.
2. ABROAD.
3. THE PHILOSOPHY OF DINNERS.
4. THE BALL.
5. THE PHILOSOPHY OF DRESS.
6. THE ART OF CONVERSATION.

CHAP.
7. THE ETIQUETTE OF WEDDINGS.
8. AT COURT.
9. HINTS ABOUT TITLES.
10. A HEALTHY LIFE.
11. TWO CENTURIES OF MAXIMS UPON MANNERS.
12. THE HOUSEHOLD.

"The most sensible book on etiquette that we remember to have seen."—*Pall Mall Gazette.*
"This book may be considered a new departure in the class of works to which it belongs. It treats etiquette 'from a liberal point of view,' and amply fulfils its purpose."—*Cassell's Papers.*
"Useful, sensibly written, and full of amusing illustrative anecdotes."—*Morning Post.*
"Creditable to the good sense and taste, as well as to the special information of its author."—*Telegraph.*
"The book is the best of the kind yet produced, and no purchaser of it will regret his investment."—*Bristol Mercury.*
"Those who live in dread lest they should not do the 'correct thing' should procure the book, which is a complete society encyclopædia."—*Glasgow News.*

WITH EIGHT ILLUSTRATIONS ON TONED PAPER.
Second edition, small crown 8vo., 352 pp., cloth, price 3s. 6d. ; gilt edges, 4s.

Girls and their Ways : A Book for and
about Girls. By ONE WHO KNOWS THEM.

CONTENTS :

CHAP.
1. THE GIRL AT HOME.
2. THE GIRL IN HER LEISURE HOURS.
3. THE GIRL AT SCHOOL—THE GIRL AND HER FRIENDS.
4. THE GIRL ABROAD : CHARACTER SKETCHES.
5. A GIRL'S GARDEN : IN PROSE AND POETRY.
6. THE GIRL'S AMATEUR GARDENER'S CALEN-

CHAP.
DAR ; OR, ALL THE YEAR ROUND IN THE GIRL'S GARDEN.
7. THE GIRL'S LIBRARY—WHAT TO READ.
8. THE GIRL IN THE COUNTRY—PASTIME FOR LEISURE HOURS THROUGHOUT THE YEAR.
9. WHAT THE GIRL MIGHT AND SHOULD BE : EXAMPLES OF NOBLE GIRLS FROM THE LIVES OF NOBLE WOMEN.

"It aims high, and it hits the mark."—*Literary World.*
"Books prepared for girls are too often so weak and twaddly as to be an insult to the intellect of girlhood. This new work is an exception."—*Daily Review (Edinburgh).*
"Worthy of a somewhat longer analysis than we shall be able to give it. . . . . Parents will be benefited by its perusal as well as their daughters. . . . . the more so that it is not written in a dry homiletic style, but with a living kindness and sympathy."—*Queen.*
"A long list of books is given both for study and amusement. This list is selected with care and without prejudice, and should prove a great assistance to girls in doubt what to read. . . . . It is a sensible and well-written book, full of information and wholesome thoughts for and about girls."—*St. James's Budget.*
"Home duties, amusement, social claims and appropriate literature, are subjects successively treated, and treated with both knowledge and sound judgment."—*Pall Mall Gazette.*

London : John Hogg, 13, Paternoster Row, E.C.

Southey's Edition, with Life of Bunyan, &c.
Illustrated with the Original Wood Blocks by W HARVEY.
Large crown 8vo., 402 pp., cloth, price 3s. 6d. ; gilt edges, 4s.

The Pilgrim's Progress. In Two

Parts. By JOHN BUNYAN. With Bibliographical Notes, and a
Life of the Author, by ROBERT SOUTHEY; Portrait and Auto-
graph of BUNYAN, and Thirty Wood Engravings by W. HARVEY,
from the Original Blocks. The Text in large type (Small Pica).
This is a reprint (with additional notes) of the deservedly admired
edition of Bunyan's Immortal Allegory, published by John Major,
London, 1830, at 21s., which was highly eulogized by Sir Walter
Scott and Lord Macaulay.

"This reprint, at a very moderate price, may be regarded as a popular boon."—*Daily Telegraph.*
"An excellent edition of the great allegory. It contains Southey's 'Life,' which certainly stands first for literary merit."—*Pall Mall Gazette.*
"Costlier editions are on sale, but none produced with more taste than this one."—*Dispatch.*
"A real service has been rendered for those who want a thoroughly readable copy of 'The Pilgrim's Progress.'"—*Literary World.*
"The whole book is reproduced in excellent fashion."—*Scotsman.*
"This edition has exceptional claims upon public favour. The late poet laureate's biography is in his best manner, while Harvey's effective woodcuts are in themselves a feature of very considerable interest to lovers of British art. In the matter of typography and general get-up the reprint is in every respect superior to the original edition, and the low price at which the book is published should tempt many to obtain a copy. The binding and decorations are very effective, and the volume is fitted to grace any drawing-room table."—*Oxford Times.*

Second Edition, with Eight Engravings after Celebrated Painters.
Small crown 8vo., 392 pp., cloth, price 3s. 6d. ; gilt edges, 4s.

The Church Seasons, Historically

and Poetically Illustrated. By ALEXANDER H. GRANT, M.A.,
Author of "Half-Hours with our Sacred Poets."

☞ The aim has been to trace the origin and history of the Festivals and Fasts of the Ecclesiastical Year, and to illustrate in poetry the circumstances under which they began and continue to be celebrated, and the principal ideas and doctrines which they severally incorporate.

"Our festival year is a bulwark of orthodoxy as real as our confessions of faith."—PRO-FESSOR ARCHER BUTLER.

"Mr. Grant's scholarship is endorsed by authorities; his method is good, his style clear, and his treatment so impartial that his work is praised alike by *Church Times, Record, Watchman, Freeman,* and *Nonconformist.* No words of ours could better prove the catholicity of a most instructive and valuable work."—*Peterborough Advertiser.*
"The work shows very plainly that much care and judgment has been used in its compilation. . . . . The intrinsic worth of its contents and their lasting usefulness admirably adapt it for a present. The eight engravings have been chosen so as to give examples of the highest samples of sacred art."—*Oxford Times.*
"A very delightful volume for Sunday reading, the devotional character of the hymns giving an especial charm to the work. The historical information will be proved full of interest to young Churchmen, and young ladies especially will find the work to be one well adapted to inform the mind and gladden the heart."—*Bible Christian Magazine.*
"Mr. Grant's volume is worthy of high praise, alike for its careful research and its discriminative quotations. There is so much religious literature which is below the level of criticism, that we cannot but welcome a volume which commends itself to a cultivated Christian audience."—*Echo.*

London: John Hogg, 13, Paternoster Row, E.C.

WITH EIGHT ILLUSTRATIONS ON TONED PAPER.

Small crown 8vo., 384 pp., cloth, price 3s. 6d. ; gilt edges, 4s.

Stories of Young Adventurers. By

ASCOTT R. HOPE, Author of "Stories of Whitminster," "A Book of Boyhoods," &c., &c.

CONTENTS :

A YOUNG TURK.
A WHITE INDIAN.
A SLAVE BOY'S STORY.
A SOLDIER BOY'S STORY.
A SAILOR BOY'S STORY.
A YOUNG YANKEE ON THE WAR PATH.
FOUR SONS OF ALBION.
A GIRL'S STORY.
AN ADVENTURER AT THE ANTIPODES.
AN ADVENTURER AT HOME.

"Mr. Hope is one of the best of living writers of boys' books, and we do not think we over-estimate the merits of the book before us if we say it is one of his best. The idea is a happy one. . . . . The result is altogether as successful as the idea is happy. This ought to be one of the most popular boys' books of the season."—*Birmingham Daily Post*.

"Good, wholesome, stirring reading for boys of all ages. The scenes of these adventures are laid in every quarter of the globe, and they include every variety of peril."—*World*.

"Mr. Ascott Hope has hit upon a really excellent idea in his 'Stories of Young Adventurers,' and carried it out with admirable success. . . . . It would be difficult to pick out a better book of its kind; young readers will hang over every page with an absorbing interest, and all the time will be imbibing some useful historical information. We should like to think that so thoroughly good a book will be in the hands of a great many boyish readers."—*Guardian*.

"Mr. Ascott Hope has won an enviable reputation as an author of books for boys. In the present volume he surpasses all his former achievements in this line."—*Literary World*.

"The book contains a great deal of good reading of a kind far superior to that which is ordinarily found in similar books. . . . . It is well got up, and will be prized by boys."—*Scotsman*.

"Sure to make the eyes of our boys gleam. . . . . The tone is healthy and robust, and for its kind the book is one of the best we know."—*Sword and Trowel*.

"A debt of gratitude is due to Mr. Hope. . . . . The work is as good as the design."—*Athenæum*.

WITH EIGHT ILLUSTRATIONS ON TONED PAPER.

Small crown 8vo., 384 pp., cloth, price 3s. 6d. ; gilt edges, 4s.

Exemplary Women : A Record of

Feminine Virtues and Achievements (abridged from "Woman's Work and Worth "). By W. H. DAVENPORT ADAMS.

CONTENTS :

CHAP.
I. WOMAN AS MOTHER.
II. WOMAN AS WIFE.
III. WOMAN AS MAIDEN.
IV. WOMAN IN THE WORLD OF LETTERS.
V. WOMAN IN THE WORLD OF ART.
VI. WOMAN AS THE HEROINE, ENTHUSIAST, AND SOCIAL REFORMER.

"The qualifications and influence of women in different spheres of life are detailed and illustrated by notices of the lives of many who have been distinguished in various positions."—*Bazaar*.

London : John Hogg, 13, Paternoster Row, E.C.

WITH EIGHT ILLUSTRATIONS ON TONED PAPER.

Small crown 8vo., 384 pp., cloth, price 3s. 6d. ; gilt edges, 4s.

A Book of Boyhoods. By Ascott R.

HOPE, Author of "Stories of Whitminster," "Our Homemade Stories," etc.

A NEW ENGLAND BOY.	A SCHOOLBOY OF THE OLDEN TIME.	A REBEL BOY.
A BRAVE BOY.	A BLUECOAT BOY.	A MYSTERIOUS BOY.
A FRENCH SCHOOLBOY.	A STABLE BOY.	A BLIND BOY.

"Well planned, well written, and well named. . . . Mr. Hope has told these stories with much dramatic power and effect, and has produced a book which will delight all healthy-minded lads."—*Scotsman.*

"Stories of all sorts of boys, who in different countries and circumstances, in peace or in war, at school or at work, at home or out in the world, by land or by sea, have gone through experiences worth relating. . . . The work is just such a volume as we would like to see in the hands of our schoolboys, and of those who are emerging into the busy haunts of business and anxiety."—*Yorkshire Gazette.*

"Mr. Ascott R. Hope now occupies the foremost place as a writer of fiction for the school boy, and as he never produces a weak book, and never disappoints his clients, his name on the title-page of a new book is always a sufficient passport. . . . The account of these young heroes is related in the happiest vein—in a style that is in itself a wholesome form of culture to the young reader. But the crowning merit of the book is that it is always interesting, and never for a moment dull."—*School Board Chronicle.*

"Essentially of an attractive character to the youthful reader, and is, perhaps, as likely to interest the sisters as the brothers."—*Bedford Mercury.*

"Ascott R. Hope has the talent for writing books which will interest boys. The volume is got up with great taste, as all Mr. Hogg's books are, and is well illustrated. A better present could not be given to a boy than this book."—*Dundee Courier.*

WITH TWELVE ILLUSTRATIONS BY THOMAS STOTHARD, R.A.,
AND A PORTRAIT OF DEFOE.

In one volume, 512 pp., large crown 8vo., cloth, price 3s. 6d. ; gilt edges, 4s.

The Life and Adventures of Robinson

Crusoe, of York, Mariner. With an Account of his Travels round Three Parts of the Globe.

☞ A Complete, unabridged Edition of both Parts, with no curtailment of the "Farther Adventures."

WITH EIGHT ILLUSTRATIONS ON TONED PAPER.

Small crown 8vo., 384 pp., cloth, price 3s. 6d. ; gilt edges, 4s.

The Ocean Wave : Narratives of some

of the Greatest Voyages, Seamen, Discoveries, Shipwrecks, and Mutinies of the World. By HENRY STEWART, Author of " Our Redcoats and Bluejackets," etc.

CONTENTS :

CHAP.
1. THE GREAT DISCOVERERS.
2. THE OLD ENGLISH SEA-KINGS.
3. THE BUCCANEERS AND THE PIRATES.
4. COMMODORE ANSON'S VOYAGE ROUND THE WORLD.
5. ADVENTURES AT SEA.
6. CAPTAIN COOK'S VOYAGES.
7. MUTINIES OF THE BRITISH NAVY.

CHAP.
8. ANECDOTES OF ENGLISH ADMIRALS FROM BLAKE TO NELSON.
9. LORD COCHRANE'S EXPLOITS.
10. STIRRING EPISODES IN THE AMERICAN CIVIL WAR.
11. ARCTIC EXPLORATION.
12. SHIPWRECKS OF RECENT TIMES.

London : John Hogg, 13, Paternoster Row, E.C.

A HANDBOOK OF REFERENCE AND QUOTATION.
Second edition, fcap. 8vo., cloth, price 2s. 6d.

Mottoes and Aphorisms from Shake-
speare : Alphabetically arranged ; with a Copious Index of 9,000 References to the infinitively varied Words and Ideas of the Mottoes. Any word or idea can be traced at once, and the correct quotation (with name of play, act, and scene) had without going further.

"The collection is, we believe, unique of its kind. . . . . It solves in a moment the often difficult question of where a proverb, or aphorism, or quotation from Shakespeare can be found."—*Oxford Times.*

"As neat a casket of Shakespearian gems as we ever remember having met with."—*Public Opinion.*

"The writer who delights now and then to embellish his productions by some of the well-pointed and telling mottoes and aphorisms from Shakespeare has here a most valuable book of reference. . . . . The work has been carefully executed, and must have entailed a very large amount of assiduous labour."—*Yorkshire Gazette.*

"Everything, in these cases, depends on the index, and the index here seems to have been carefully made."—*Sheffield Independent.*

New and enlarged edition, fcap. 8vo., cloth, price 2s. 6d.

A Practical Guide to English Versifi-
cation, with a Compendious Dictionary of Rhymes, an Examination of Classical Measures, and Comments upon Burlesque and Comic Verse, Vers de Société, and Song Writing. By TOM HOOD. A new and enlarged edition, to which are added Bysshe's " RULES FOR MAKING ENGLISH VERSE," etc.

"We do not hesitate to say, that Mr. Hood's volume is deserving of a place on the shelves of all who take an interest in the structure of verse."—*Daily News.*

"The book is compiled with great care, and will serve the purpose for which it is designed. . . . . We may add that it contains a good deal of information which will be useful to students who have no wish to be numbered amongst verse-makers."—*Pall Mall Gazette.*

"A dainty little book on English verse-making. The Dictionary of Rhymes will be found one of the most complete and practical in our language."—*Freeman.*

"Alike to the tyro in versifying, the student of literature, and the general reader, this guide can be confidently recommended."—*Scotsman.*

Crown 8vo., cloth extra, bevelled boards, price 7s. 6d.

The Manuale Clericorum : A Guide
for the Reverent and Decent Celebration of Divine Service, the Holy Sacraments, and other Offices, according to the Rites, Ceremonies, and Ancient Use of the United Church of England and Ireland. Abridged from the " Directorium Anglicanum." With Additions of Special Value in the Practical Rendering of the Services of the Church. Edited by the Rev. F. G. LEE, D.C.L., F.S.A., Vicar of All Saints', Lambeth.

London : John Hogg, 13, Paternoster Row, E.C.

Demy 8vo., 792 pages, price 15s.

Dedicated by permission to the late JOHN HERVEY, Esq., Grand Secretary.

The Royal Masonic Cyclopædia of
History, Rites, Symbolism, and Biography.   Containing upwards
of 3,000 Subjects, together with numerous Original Articles on
Archæological and other Topics.   Edited by KENNETH R. H.
MACKENZIE, IX°.

" The work is marked by extreme learning and moderation."—*Public Opinion.*
" We welcome this laborious work very sincerely."—*Freemason.*
" A really valuable and instructive work, alike interesting to the Masonic Student and
general reader, and to the curious it will prove to be an inexhaustible mine of wealth,
particulars being afforded of numerous strange subjects. . . . Deserves a large circulation,
and cannot fail to be a most welcome work in every Masonic library."—*Keystone.*
" The most valuable work of reference on all matters relating to the Craft that has yet been
published."—*British Mercantile Gazette.*
" The task has been admirably performed. . . . . One of the most important additions
to Masonic Literature during the last quarter of a century, and deserves an honoured place in
the library of every Masonic Student."—*Freemason's Chronicle.*
" The Editor has lavished much reading and labour on his subject."—*Sunday Times.*

Crown 8vo., cloth, with diagrams, price 4s. 6d.

The Discrepancies of Freemasonry;
Examined during a Week's Gossip with the late celebrated Bro.
Gilkes, and other Eminent Masons.   By the late Rev. G.
OLIVER, D.D.

" It is difficult to imagine a more charming book, or one more calculated to inspire the
Masonic Student with enthusiasm for the Royal Art.   The pen of a practical as well as a
ready writer is needed in writing dialogues, and . . . . the late Sir Arthur Helps is the
only man of eminence who could possibly have infused more interest into such a work."—
Freemason's Chronicle.
" A most amusing and curious book."—*Standard.*

Crown 8vo., cloth, with diagrams, price 4s. 6d.

The Pythagorean Triangle ; or, the
Science of Numbers.   By the late Rev. G. OLIVER, D.D.

" In addition to all its stores of curious and varied learning, as connected with the Craft,
the Rev. Doctor's treatise contains many sage remarks on a host of other interesting topics,
which will please all curious readers."—*Standard.*
" In handling his subject, the author has shown even more than his usual skill and ingenuity."—*Freemason's Chronicle.*
" We have derived both information and entertainment from this volume."—*Literary
World.*
" From first chapter to last it will be impossible to read a more interesting book, illustrative
of the symbolism of Freemasonry."—*British Mail.*

Second edition, demy 8vo., 56 pages, price 8d.

Freemasonry: Its History, Principles,
and Objects.

" We most cordially recommend this little work to the serious perusal, not only of those
who are already numbered amongst the Craft, but also of all who may meditate on entering
the ranks of Freemasonry.   It is a *vade mecum* of very convenient form, and although consisting of only fifty-six pages, the amount of Masonic lore therein contained is really astonishing."—*Sunday Times.*

London : John Hogg, 13, Paternoster Row, E.O.

Pocket size, cloth, gilt edges, price 2s. 6d. each.

Masonic Directories. A Series of

Four Handbooks of Practical Directions for the Efficient Conduct of the Work throughout the Three Degrees of Craft Masonry. By KENNETH R. H. MACKENZIE, IX° ("Cryptonymus"), Author of "The Royal Masonic Cyclopædia," etc.

I. The Deacons' Work.
II. The Wardens' Work.
III. The Secretary's and Treasurer's Work.
IV. The W. Master's Work.

WHAT IS SAID ABOUT THE "MASONIC DIRECTORIES."

The following spontaneous expression of opinion from one of the Craft, who had ordered the "Directories," is indicative of the favourable reception which the books have met with on all hands :—

"It is simply impossible to speak too highly of these little books, being well put together, simple, perfect, and yet within the reach of all. The four Directories supply a want long felt. Every Master of a Lodge should order a supply of these Directories, and bring them seriously to the notice of the Officers for whom they are intended. If this were done, I have no doubt that the great drawback which exists in very many Lodges, resulting from the fact of Deacons, Wardens, Treasurer, Secretary, and I regret to say occasionally the W.M., not being well up in their duties, might be remedied. The Author of the 'Royal Masonic Cyclopædia' is deservingly entitled to the grateful thanks of every true Mason for his labours in Masonic writing."

And as representative of Press Criticism, what the *Yorkshire Gazette* said the other day may be cited :—" We do not hesitate to recommend them to members of the Craft. They are very reliable, and are printed in a neat and handy form. We suspect that there are few working members of our Order who would not be benefitted by the results of Brother Mackenzie's observations and experience."

THIRD AND CHEAPER EDITION, REVISED AND ENLARGED.
Crown 8vo., cloth, with 14 illustrations, price 7s. 6d.

The Freemason's Manual; or, Illus-

trations of Masonry. By JEREMIAH HOW, K.T., 30°, P.M., P.Z., etc.

Imperial 16mo., with a frontispiece, cloth, marbled edges, price 7s. 6d.

The Complete Manual of Oddfellow-

ship : Being a Practical Guide to its History, Principles, Ceremonies, and Symbolism.

The Ritual is printed in a form intelligible only to the Order.

London : John Hogg, 13, Paternoster Row, E.C.

CLASSIFIED CONTENTS OF CATALOGUE.

———◦◊◦———

London : John Hogg, 13, Paternoster Row, E.C.

www.ingramcontent.com/pod-product-compliance
Lightning Source LLC
Chambersburg PA
CBHW021508210326
41599CB00012B/1171